PERGAMON INTERNATIONAL LIBRARY
of Science, Technology, Engineering and Social Studies

*The 1000-volume original paperback library in aid of education,
industrial training and the enjoyment of leisure*

Publisher: Robert Maxwell, M.C.

HUMAN
RELIABILITY

THE PERGAMON TEXTBOOK
INSPECTION COPY SERVICE

An inspection copy of any book published in the Pergamon International Library
will gladly be sent to academic staff without obligation for their consideration for
course adoption or recommendation. Copies may be retained for a period of 60 days
from receipt and returned if not suitable. When a particular title is adopted or
recommended for adoption for class use and the recommendation results in a sale
of 12 or more copies the inspection copy may be retained with our compliments.
The Publishers will be pleased to receive suggestions for revised editions and new
titles to be published in this important international Library.

Pergamon Titles of Related Interest

Baylis SAFETY OF COMPUTER CONTROL SYSTEMS

Bushnell TRAINING FOR NEW TECHNOLOGY

Guzzo A GUIDE TO WORKER PRODUCTIVITY EXPERIMENTS
IN THE UNITED STATES, 1976–81

Guzzo PROGRAMS FOR PRODUCTIVITY & QUALITY
OF WORKING LIFE

Gvishiani SYSTEMS RESEARCH

ICE LOSS PREVENTION & SAFETY PROMOTION IN
THE PROCESS INDUSTRIES

ICE RELIABLE PRODUCTION IN THE PROCESS INDUSTRIES

ICE THE ASSESSMENT OF MAJOR HAZARDS

Johannsen & Rijnsdorp ANALYSIS, DESIGN & EVALUATION
OF MAN-MACHINE SYSTEMS

Martin DESIGN OF WORK IN AUTOMATED
MANUFACTURING SYSTEMS

O'Brien et al. INDUSTRIAL BEHAVIOR MODIFICATION

Ozawa PEOPLE AND PRODUCTIVITY IN JAPAN

Ricci & Rowe HEALTH & ENVIRONMENTAL RISK ASSESSMENT

Sanchez & Gupta FUZZY INFORMATION, KNOWLEDGE
REPRESENTATION & DECISION ANALYSIS

Tomlinson RETHINKING THE PROCESS OF OPERATIONAL
RESEARCH & SYSTEMS ANALYSIS

Related Journals
(Sample copies available on request)

ACCIDENT ANALYSIS & PREVENTION
COMPUTERS AND INDUSTRIAL ENGINEERING
COMPUTERS AND OPERATIONS RESEARCH
COMPUTERS IN HUMAN BEHAVIOR
INTERNATIONAL JOURNAL OF FORENSIC ENGINEERING
INTERNATIONAL JOURNAL OF HUMAN RELIABILITY
JOURNAL OF BIOMECHANICS
JOURNAL OF PRODUCTS LIABILITY
JOURNAL OF SAFETY RESEARCH
MATHEMATICAL MODELLING
MICROELECTRONICS AND RELIABILITY
NEW IDEAS IN PSYCHOLOGY
SYSTEMS RESEARCH
WORK IN AMERICA INSTITUTE STUDIES IN PRODUCTIVITY

HUMAN
RELIABILITY
With Human Factors

Balbir S. Dhillon

Department of Mechanical Engineering
University of Ottawa

PERGAMON PRESS

New York Oxford Beijing Frankfurt São Paulo Sydney Tokyo Toronto

Pergamon Press Offices:

U.S.A.	Pergamon Press, Maxwell House, Fairview Park, Elmsford, New York 10523, U.S.A.
U.K.	Pergamon Press, Headington Hill Hall, Oxford OX3 0BW, England
PEOPLE'S REPUBLIC OF CHINA	Pergamon Press, Qianmen Hotel, Beijing, People's Republic of China
FEDERAL REPUBLIC OF GERMANY	Pergamon Press, Hammerweg 6, D-6242 Kronberg-Taunus, Federal Republic of Germany
BRAZIL	Pergamon Editora, Rua Eça de Queiros, 346, CEP 04011, São Paulo, Brazil
AUSTRALIA	Pergamon Press (Aust.) Pty., P.O. Box 544, Potts Point, NSW 2011, Australia
JAPAN	Pergamon Press, 8th Floor, Matsuoka Central Building, 1-7-1 Nishishinjuku, Shinjuku, Tokyo 160, Japan
CANADA	Pergamon Press Canada, Suite 104, 150 Consumers Road, Willowdale, Ontario M2J 1P9, Canada

First printing 1986

Library of Congress Cataloging in Publication Data

Dhillon, B.S.
  Human reliability.

  Includes bibliographical references and index.
    1. Human engineering. 2. Reliability (Engineering)
I. Title.
TA166.D49 1986   620.8′2   86-2389
ISBN 0-08-032774-5
ISBN 0-08-033981-6 (pbk.)

620.82
D534h

Printed in Great Britain by A. Wheaton & Co. Ltd., Exeter

This book is affectionately dedicated
to my grandmother, Bichint Kaur

Contents

List of Figures

List of Tables

Preface and Acknowledgments

Modern technology has created a tendency to produce equipment and systems of greater capital cost, sophistication, complexity and capacity. The consequences of unreliable behavior of equipment and systems have become increasingly more severe and have led to the desire for better reliability. Nowadays complex system reliability analysis is no longer restricted to the hardware aspect only, but also takes into consideration other aspects, such as reliability of the human element and that of the software. Although it was during the years of World War II that human factors began to be regarded as a somewhat distinct discipline, it was not until the late 1950s that it was clearly stated that realistic system reliability analysis must include the human aspect.

Ever since the beginning of the 1960s there has been considerable growth of the published literature on the topic of human reliability. Interest in this subject has revived further because the well-publicized Three Mile Island nuclear accident was the result of hardware failures and human error.

Nowadays there is a tendency to place greater emphasis on human reliability during system design. A technical professional faces inconvenience in securing information on the subject of human reliability and related areas because the information is covered in technical reviews or only briefly in some textbooks, but (to the author's best knowledge) not specifically in a single volume. This book is an attempt to fulfill this vital need. It emphasizes concepts and avoids getting bogged down in mathematical rigor and details. Despite this, it holds great utility for persons with engineering backgrounds. At the end of each chapter, the sources of most of the materials presented are listed. This will provide a useful service to readers for further investigation, whenever it is necessary. Several examples, along with their corresponding solutions, are presented in the text to help in understanding its contents. Although the prime objective of the book is to cover human reliability, nevertheless some of the related areas are also discussed. Understanding of such areas is also quite useful in human reliability work. This text will be useful to readers such as human factor engineers and specialists, reliability and maintainability specialists, system and design engineers, industrial engineers, quality control engineers and students.

The book is composed of 13 chapters. Chapter 1 briefly discusses the histories of human factors and human reliability along with selective terms and definitions. A review of the basic reliability mathematics and concepts helpful in understanding the contents of subsequent chapters is presented in chapter 2. Chapter 3 introduces the topic of human reliability. The topics of stress, mathematical modeling of human error occurrence, human performance reliability modeling in a continuous time domain and the fault-tree method are covered.

Chapter 4 is completely devoted to human errors. Various aspects of human errors are discussed. Some of these are the classifications of human errors, reasons for human errors and human-error prevention methods. Six human reliability analysis methods are presented in chapter 5.

Chapter 6 is concerned with the reliability evaluation of systems with human errors. It contains several Markov models. The theme of chapter 7 is human factors in maintenance and maintainability. Important aspects of both these topics are discussed. Chapter 8 deals with the important topic of human safety. Some of the subjects covered in the chapter are accident losses, reasons for accidents, accident- and error-reduction measures, safety devices and human failure modes.

The important topic of human reliability data is discussed in chapter 9. This chapter addresses various important areas of the human reliability data. Human factors in quality control is the theme of chapter 10. The material covered in this chapter is concerned with management and operator-controllable errors, inspector errors, inspection-related mathematical models, and so on.

Chapters 11–13 present three significant areas related to human factors, i.e., human factors in design, mathematical models and formulas and applications of human factors engineering. The material discussed in these three chapters is considered to be of significance and its knowledge is essential in human reliability work.

The author wishes to thank many friends, colleagues and leading professionals who, through discussions, have shaped his thinking on several areas of this book. In particular, I am grateful to Dr. K. B. Klaassen of IBM and Mr. Thomas Anthony of Pergamon Press for their useful comments on the first draft of this book. I am deeply indebted to Dr. S. N. Rayapati for preparing all the diagrams for this text. I wish to express my thanks to my parents, brother, relatives and friends for their interest and encouragement at the moments of need. Finally, I thank my wife, Rosy, for typing the entire manuscript and proofreading. Her patience and tolerance have helped quite a lot during the preparation of the manuscript!

Balbir S. Dhillon
Ottawa, Ontario, Canada

Chapter 1

Introduction

In recent years increasing attention is being paid to human reliability. This is witnessed by the increasing number of publications on the subject. There are various reasons for this increase. One reason could be that engineering systems have become highly sophisticated and complex. Moreover, failure of such systems may produce far-reaching and unpredictable effects. The prime example of complex systems failure is the Three Mile Island nuclear accident. This accident was the result of a combination of human error and hardware failure. Another factor for the increasing attention paid to human reliability could be that various studies have indicated that a significant proportion of system failures are due to human errors. According to Ref. [1], about 50–70% of the failures in electronic equipment were human-initiated, whereas in aircraft and missile systems the human-initiated failures accounted for 60–70% and 20–53% of the total failures, respectively. Furthermore, in Ref. [2] it is stated that about 10–15% of the total failures were directly due to humans.

Nowadays, the effort is directed toward replacing the human functions with machines and monitoring the human with computers. The main objective of this effort is to reduce the occurrence of human errors. Even highly automated systems do not totally remove human involvement. Therefore, it is not wrong to state that without giving proper consideration to human reliability during the system design phase, the reliability analysis will be incomplete.

HISTORY OF HUMAN FACTORS

The history of human factors engineering may be traced back as early as the earliest human. For example, pebble tools were used by *Australopithecus*

1

prometheus, who in addition employed thigh bones as weapons [3,4]. However, in modern times Frederick W. Taylor would probably be called the first human factors engineer. In 1898, he conducted studies to find the most appropriate designs for shovels [5]. In 1911, Frank B. Gilbreth conducted a study of bricklaying. This resulted in the invention of a scaffold. This scaffold enabled bricklayers to work at the most suitable level at all times because it could be raised or lowered quickly. With the result of the Gilbreth study, the bricklayers were able to lay bricks at a rate of 120–350 per man-hour. This represented a significant increase in the bricklayers' output.

During World War I, the governments of the United States and of the United Kingdom directed significant attention to military personnel selection and training. The prime target of this effort was "fitting the man to the job." In 1918, in the United States, laboratories were established at the Wright–Patterson Air Force Base and the Brooks Air Force Base to perform human-factors-related research [6]. These laboratories have performed research on areas such as complex reaction time, perception and motor behavior.

The years between the two World Wars witnessed major growth in disciplines such as industrial psychology and industrial engineering. During World War II engineering systems became so complex and sophisticated that the need for human factors consideration became mandatory.

By 1945 human factors engineering as a specialized discipline was recognized. In the 1950s and 1960s the military and manned space programs further increased the importance of human factors. At present, several textbooks on the subject have appeared and a number of research journals are devoted to this field.

BRIEF HISTORY OF HUMAN RELIABILITY

Since World War II reliability engineering has been receiving considerable attention. Five research journals in English alone are fully or partially devoted to this discipline. Over 100 books on reliability have been published since the late 1950s.

In 1958, H. L. Williams [7] was one of the first persons who recognized that human-element reliability must be included in the system-reliability prediction; otherwise the predicted system reliability would not represent the real picture. Two years later, in 1960, A. Shapero et al. [8] pointed out that human error is the cause for a large proportion (i.e., from 20 to 50%) of all equipment failures. In the same year, W. I. LeVan [9] reported that 23–45% of the failures resulted from human error. In the 1960s, a number of publications related to human reliability appeared in journals, conference proceedings and technical reports. Many of these publications are listed in Ref. [10]. Two of the important documents of the 1960s were the *Proceedings of the*

Symposium on the Reliability of Human Performance in Work [11] and the *Proceedings of the Symposium on Quantification of Human Performance* [12]. The year 1973 may be regarded as an important milestone in the history of human reliability. It was in August of that year when a well-known journal entitled *IEEE Transactions on Reliability* [13] published a special issue devoted to human reliability. A number of excellent papers appeared in this issue. Ever since, researchers have been making further advances in the human reliability field. A selective bibliography on human reliability is presented in Ref. [14]. This bibliography covers the period from 1958 to 1978.

TERMS AND DEFINITIONS

This section presents selective terms and definitions used in human or general reliability [2,15,16].

Human reliability. This is the probability of accomplishing a job or task successfully by humans at any required stage in system operation within a specified minimum time limit (if the time requirement is specified).

Human error. This is the failure to carry out a specified task (or the performance of a forbidden action) that could lead to disruption of scheduled operations or result in damage to property and equipment.

Human factors. This is a body of scientific facts concerning the characteristics of human beings. The term embraces all biomedical and psychosocial considerations. It includes, but is in no way restricted to, personnel selection, training principles and applications in the area of human engineering, evaluation of human performance, aids for job performance and life support.

Human engineering. This is the area of human factors considerations that makes use of scientific facts in the design of items to produce effective man-machine integration and utilization effectively.

Reliability. This is the probability that an item will perform its specified function for a stated time under specified conditions.

Availability. This is the probability that an item is available for use when required.

Continuous task. This is a task involving some kind of tracking activity (one example of such activity is monitoring a changing situation).

Human performance reliability. This is the probability that a human will fulfill all specified human functions subject to stated conditions.

Steady–state condition (statistical). This is that condition where the probability of being in a particular state does not depend on time.

Redundancy. This is the existence of two or more means for completing a specified function.

Man-function. That function which is allocated to the system's human element.

Human performance. This is a measure of man-functions and actions under stated conditions.

SCOPE OF THE BOOK

In recent years increasing attention has been paid to human reliability during the design, manufacturing and operation phases of engineering systems. The topic of human reliability is discussed in various technical papers, reports and specialized books, and at present to the best of the author's knowledge is not covered within the framework of a single book. Engineers and others needing detailed information on human reliability and related areas face a tremendous inconvenience. This book is an attempt to fulfill this vital need because there has been a considerable growth of the knowledge concerning human reliability. The book is written in such a way that previous knowledge is not necessary to digest its contents. For example, chapter 2 presents basic mathematics and reliability theory. This eliminates the need for prior knowledge of basic mathematics and reliability. In addition, many examples, presented with solutions, make this text more self-explanatory.

This book will be applicable across many disciplines because a common problem is human error. It will be useful to reliability and maintainability engineers and specialists; human factors engineers and specialists; design and systems engineers; quality control engineers; industrial engineers; electrical, mechanical, civil and chemical engineers; and senior undergraduate and graduate students.

SUMMARY

This chapter briefly discussed the histories of human factors engineering and human reliability. The importance of human reliability is briefly described. Selective definitions and terms related to human and general reliability are presented. The scope of the text is briefly outlined.

EXERCISES

1. Describe the present trend in the growth of human reliability field.
2. Define the following terms:
   a. Stress (human)
   b. Human error rate.
   c. System steady-state unavailability.
3. Discuss the histories of reliability and maintainability engineering.

REFERENCES

1. J. M. Christensen, J. M. Howard and B. S. Stevens, Field experience in maintenance, in *Human Detection and Diagnosis of System Failures* (edited by J. Rasmussen and W. B. Rouse), pp. 111–133, Plenum Press, New York (1981).
2. E. W. Hagen (Ed.), Human reliability analysis. *Nuclear Safety*, **17**, 315–326 (1976).
3. R. A. Dart, *Adventures with the Missing Link*. The Viking Press, New York (1959).
4. J. M. Christensen, The evolution of the systems approach in human factors engineering, in *Proceedings of the University of Michigan Human Factors Engineering Summer Conference*, University of Michigan, Ann Arbor, Michigan, pp. 29.1–29.10 (1964).
5. A. Chapanis, *Man–Machine Engineering*. Wadsworth Publishing Company, Inc., Belmont, California (1965).
6. D. Meister and G. F. Rabideau, *Human Factors Evaluation in System Development*. John Wiley & Sons, New York (1965).
7. H. L. Williams, Reliability evaluation of the human component in man–machine systems. *Electrical Manufacturing*, 78–82 (April 1958).
8. A. Shapero, J. I. Cooper, M. Rappaport, K. H. Shaeffer and C. J. Bates, *Human Engineering Testing and Malfunction Data Collection in Weapon System Programs*. WADD Technical Report, 60-36, Wright–Patterson Air Force Base, Dayton, Ohio (February 1960).
9. W. I. LeVan, *Analysis of the Human Error Problem in the Field*. Report No. 7-60-932004, Bell Aerosystems Company, Buffalo, New York (June 1960).
10. D. Meister, *Human Factors: Theory and Practice*, pp. 54–56. John Wiley & Sons, New York (1971).
11. W. B. Askren (Ed.), *Proceedings of the Symposium on the Reliability of Human Performance in Work*. Report AMRL-TR-67-88, Aerospace Medical Research Laboratories, Wright–Patterson Air Force Base, Ohio (May 1967).
12. *Proceedings of the Symposium on Quantification of Human Performance*, sponsored by the Electronics Industries Association and the University of New Mexico, Albuquerque, New Mexico (August 1964).
13. T. L. Regulinski (Ed.), Special issue on human reliability. *IEEE Transactions on Reliability*, **22** (August 1973).
14. B. S. Dhillon, On human reliability — bibliography. *Microelectronics and Reliability*, **20**, 371–373 (1980).
15. *Definitions of Effectiveness Terms for Reliability, Maintainability, Human Factors and Safety*; MIL-STD-721B (August 1966). Available from the Naval Publications and Forms Center, 5801 Tabor Ave., Philadelphia, PA, 19120.
16. D. Meister, Human factors in reliability, in *Reliability Handbook* (edited by W. G. Ireson), pp. 12.2–12.37, McGraw-Hill, New York (1966).

Chapter 2

Mathematics and Basic Reliability Concepts

INTRODUCTION

Just as in the development of many other disciplines, mathematics and probability theory have played an important role in the development of reliability engineering. Obviously, the history of mathematics is older than the disciplines that make use of mathematics. For example, our present number symbols can be found on stone columns erected by the famous Indian King Ashoka in about 250 BC [1]. On the other hand, the history of probability theory is not as old as that of mathematics, although the Greek philosophers of antiquity pointed out the importance of probability. However, it was not until the fifteenth century and the early years of the sixteenth century that some Italian mathematicians directed their efforts to determining the winning chances in some gambling games. For example, Girolamo Cardano (1501-1576), who held important chairs at the Universities of Bologna and Pavia, wrote a gambler's manual in which questions on probability were considered. However, it was not until the year 1654 that the probability problem was put forward to Blaise Pascal (1623-1662) by the Chevalier de Mere, an able and experienced gambler. Pascal communicated the problem to a fellow Frenchman, Pierre de Fermat (1601-1665), a great number theorist [2]. The probability problem was solved by both these men differently but correctly. Ever since those days various other researchers have contributed to the probability field.

In comparison to mathematics and probability the history of the reliability field is not that old. In the early 1930s probability concepts were applied probably for the first time to electric power generation problems [3-5]. However, the real beginning of the reliability field is usually regarded as around World War II. It was in this war that German scientists introduced and applied the basic reliability concept (a chain cannot be made stronger than its weakest link) to improve the reliability of their V1 rocket. Since the 1940s many researchers and authors have contributed to the development of the reliability field. This chapter briefly presents the various aspects of mathematics and basic reliability concepts.

PROBABILITY

Definition

Probability may simply be stated as the measure of the occurrence likelihood of an event. In mathematical terms, probability may be stated as follows: if in a series of k trials, the event E occurs Y times, and the value of Y/k for increasing k approaches a limit P, then the probability of occurrence of the event E is P. Thus we may write

$$P(E) = \lim_{k \to \infty} (Y/k) \ , \qquad (2.1)$$

where $P(E)$ is the probability of occurrence of event E. Usually the value of $P(E)$ is approximated by

$$P(E) = Y/k \ . \qquad (2.2)$$

EXAMPLE 2.1

A division of a company manufactures 1000 electric switches per month. Over the one-year period 800 switches were rejected by the quality control department. Calculate the probability of rejecting a switch being inspected.

The total number of switches manufactured in the one year time period is given by

$$k = (1000)(12)$$

$$= 12,000 \text{ switches} \ .$$

Similarly, the total number of switches rejected during the same period is

$$Y = 800 \text{ switches} \ .$$

With the aid of Eq. (2.2), the probability of rejecting a switch being inspected is

$$P = \frac{Y}{k} = \frac{800}{12,000} = 0.067 \ .$$

This means that the chances for rejecting a switch are 6.7%.

Independent Events

Two events, say, X_1 and X_2, are said to be independent if the occurrence or nonoccurrence of one event does not influence the probability of occurrence of another event. For examples, events X_1 and X_2 are independent if

$$P(X_1 \cdot X_2) = P(X_1) \cdot P(X_2) \, , \tag{2.3}$$

where $P(X_1 \cdot X_2)$ is the probability of an intersection of events X_1 and X_2, $P(X_1)$ is the probability of occurrence of event X_1, and $P(X_2)$ is the probability of occurrence of event X_2.

In the left-hand side of Eq. (2.3) the intersection of events X_1 and X_2 is denoted by a dot.

If there are m events, i.e. $X_1, X_2, X_3, X_4, \ldots$, the general form of Eq. (2.3) becomes

$$P(X_1 X_2 X_3 \ldots X_m) = P(X_1) P(X_2) P(X_3) \ldots P(X_m)$$

$$= \prod_{i=1}^{m} P(X_i) \, . \tag{2.4}$$

Example 2.2

An operator is required to perform a task composed of two independent subtasks X_1 and X_2. Each subtask is either performed correctly or incorrectly. For task success both subtasks must be performed correctly. The probabilities of performing subtasks X_1 and X_2 correctly are 0.95 and 0.8, respectively. Compute the probability of performing the task successfully by the operator.

In this example we assume that the probability, P, of performing subtask X_1 correctly is

$$P_1 = 0.95 \, ,$$

and similarly the probability P_2 of performing subtask X_2 correctly is

$$P_2 = 0.8 \, .$$

Finally, with the aid of Eq. (2.3), the probability P_s of performing the task successfully by the operator is

$$P_s = P_1 \cdot P_2$$

$$= 0.76 \, .$$

The above result indicates that if the task was composed of many subtasks with similar success probabilities, the overall result could have been very discouraging.

Conditional Probability

This is associated with a situation where the probability of occurrence of an event, say X, may depend on the occurrence or nonoccurrence of another event, say, Y. Thus the conditional probability of event X given that event Y has occurred is given by

$$P(X/Y) = \frac{P(X \cdot Y)}{P(Y)} \ , \tag{2.5}$$

where $P(X/Y)$ is the conditional probability of X given Y, $P(Y)$ is the probability of event Y, and $P(X \cdot Y)$ is the probability of the intersection of events X and Y.

EXAMPLE 2.3

A highway is 400 mi long. Anywhere on the highway the occurrence of accidents is equally likely [7]. Therefore, it is assumed that the probability of occurrence of an accident in a specified highway interval is proportional to the interval distance. Events X and Y denote an accident in 0–300 mi and in 250–350 mi, respectively. Calculate the occurrence probability of the event X if an accident occurs in the interval from 250 to 350 mi.

The probabilities of events X and Y are

$$P(X) = \frac{300-0}{400} = 0.75$$

and

$$P(Y) = \frac{350-250}{400} = 0.25 \ .$$

Similarly,

$$P(X \cdot Y) = \frac{300-250}{400} = 0.125 \ .$$

Finally, with the aid of Eq. (2.5) the occurrence probability of the event X, if an accident occurs in the interval from 250 to 350 mi, is given by

$$P(X/Y) = \frac{P(X \cdot Y)}{P(Y)} = \frac{0.125}{0.25} = 0.5 \ .$$

This means that there is only a 50% chance for the occurrence of event X.

Probability of the Union of Events

For m events this is given by

$$P(X_1 + X_2 + X_3 + \ldots + X_m) = 1 - \prod_{i=1}^{m} [1 - P(X_i)] \ , \qquad (2.6)$$

where X_i is the ith event for $i = 1, 2, 3, \ldots, m$, and $P(X_i)$ is the probability of event X_i for $i = 1, 2, 3, 4, \ldots, m$.

In the left-hand side of (2.6) the union of events is denoted by the symbol $+$.

For two independent events, i.e., $m = 2$, (2.6) simplifies to

$$\begin{aligned}
P(X_1 + X_2) &= 1 - \prod_{i=1}^{2} [1 - P(X_i)] \\
&= 1 - [1 - P(X_1)][1 - P(X_2)] \\
&= P(X_1) + P(X_2) - P(X_1) \cdot P(X_2) \ . \qquad (2.7)
\end{aligned}$$

For m mutually exclusive events, Eq. (2.6) becomes

$$P(X_1 + X_2 + X_3 + \ldots + X_m) = \sum_{i=1}^{m} P(X_i) \ . \qquad (2.8)$$

EXAMPLE 2.4

An operator performs a task composed of two independent subtasks A and B. At least one of the subtasks must be performed correctly for task success. Each of the two subtasks is either performed correctly or incorrectly. Calculate the probability of accomplishing the task correctly if the probabilities of performing subtasks A and B correctly are 0.9 and 0.95, respectively.

By substituting the specified data into Eq. (2.7) we get

$$\begin{aligned}
P(A + B) &= P(A) + P(B) - P(A) \cdot P(B) \\
&= 0.9 + 0.95 - (0.9)(0.95) \\
&= 0.995 \ .
\end{aligned}$$

Thus the probability of accomplishing the task correctly is 0.995. In comparison to the end result of Example 2.2, the chances for performing the task correctly have improved.

EXAMPLE 2.5

A fluid-flow valve may fail in either open mode or close mode. More clearly, it has two mutually exclusive failure events. The probabilities of open-mode and close-mode failures are 0.05 and 0.10, respectively. Calculate the fluid-flow-valve reliability.

We define the events

$$X_1 = \text{open-mode failure},$$

$$X_2 = \text{close-mode failure}.$$

The fluid-flow valve has only two failure modes. Thus in Eq. (2.8) the value of m is equal to 2. From Eq. (2.8), the total probability of the valve failure is

$$P(X_1 + X_2) = \sum_{i=1}^{2} P(X_i)$$

$$= P(X_1) + P(X_2)$$

$$= 0.05 + 0.10$$

$$= 0.15 .$$

By subtracting the above result from unity we get the fluid-flow-valve reliability, R_{fv}, as follows:

$$R_{fv} = 1 - P(X_1 + X_2) = 0.85 .$$

PROBABILITY DISTRIBUTIONS

This section presents two well-known probability distributions. These are the exponential and binomial distributions.

Exponential Distribution

This distribution belongs to the family of continuous random-variable functions. The exponential distribution is widely used in reliability engineering. In reliability engineering the probability density function, $f(t)$, of this distribution is defined as follows:

$$f(t) = \lambda e^{-\lambda t} \quad \text{for } \lambda > 0 \ , \quad t \geq 0 \ , \tag{2.9}$$

where λ is the constant failure rate and t is time. The cumulative distribution function, $F(t)$, for a continuous random variable T is given by

$$F(t) = \int_{-\infty}^{t} f(t) \, dt \ . \tag{2.10}$$

By subtracting the above relationship from unity we get the reliability function

$$R(t) = 1 - F(t)$$

$$= 1 - \int_{-\infty}^{t} f(t) \, dt \ . \tag{2.11}$$

The above expression is usually written as

$$R(t) = 1 - \int_{0}^{t} f(t) \, dt \tag{2.12}$$

or

$$R(t) = \int_{t}^{\infty} f(t) \, dt \ . \tag{2.13}$$

EXAMPLE 2.6

An engineering system's failure times are described by Eq. (2.9). Obtain an expression for the system reliability function.

By substituting Eq. (2.9) into (2.10) the following expression for the cumulative distribution function results:

$$F(t) = \int_{0}^{t} \lambda e^{-\lambda t} \, dt$$

$$= 1 - e^{-\lambda t} \ . \tag{2.14}$$

After subtracting Eq. (2.14) from unity, we get the following expression for the system reliability:

$$R(t) = 1 - (1 - e^{-\lambda t}) = e^{-\lambda t} \ . \tag{2.15}$$

EXAMPLE 2.7

An operator is performing a time-continuous task. The operator error rate is 0.02 error/hr. Evaluate the operator unreliability for a 3 hr mission.
In this example the following values for λ and t are specified:

$$\lambda = 0.02 \text{ error/hr} \ ,$$

$$t = 3 \text{ hr} \ .$$

Using the data in (2.14) we get the following value for the operator reliability:

$$R(3) = e^{-(0.02)(3)}$$
$$= 0.9418 \ .$$

Thus the operator unreliability, R_{ou}, is

$$R_{\text{ou}} = 1 - 0.9418$$
$$= 0.0582 \ .$$

More clearly, there is a 5.82% chance for operator error during the 3 hr mission.

Binomial Distribution

This distribution belongs to the family of discrete random-variable functions. The binomial distribution has many applications in reliability engineering. This distribution can be applied in situations where a series of trials satisfy the following three conditions:

1. Trials are independent.
2. Each trial has exactly two outcomes; for example, "success" or "failure" of an event.
3. The event occurrence probability in each trial is constant.

Thus the probability of exactly n occurrences in m trials is given by

$$P(N = n) = \left[\frac{m!}{n!(m-n)!} \right] R^n (1-R)^{m-n} \qquad \text{for } n = 0,1,2,3,\ldots m \ ,$$

$$(2.16)$$

where R is the occurrence probability of an event in each trial, $m! = 1.2.3.4..m$, and $0! = 1$.

The following result proves that the total probability P_T, is equal to unity:

$$P_T = \sum_{i=0}^{m} \left[\frac{m!}{i!\,(m-i)!} \right] R^i (1-R)^{m-i}$$

$$= [R + F]^m$$

$$= 1 , \tag{2.17}$$

where $F = 1 - R$.

EXAMPLE 2.8

A system is composed of three independent and identical units. Each unit's reliability is 0.8. Calculate the probability of exactly two units operating normally.

With the aid of Eq. (2.16) and the given data we get

$$P(N = 2) = \left[\frac{3!}{2!\,(3-2)!} \right] R^2 (1-R)^{3-2}$$

$$= 3R^2(1-R)$$

$$= 3(0.8)^2(1 - 0.8)$$

$$= 0.3840 .$$

Thus, the probability of exactly two units operating normally is 0.3840.

EXAMPLE 2.9

In Eq. (2.17) the specified values for m and R are 3 and 0.9, respectively. Prove that the value of P_T is equal to unity.

Substituting the specified values for m and R in Eq. (2.17) yields

$$P_T = \sum_{i=0}^{3} \left[\frac{3!}{i!\,(3-i)!} \right] (0.9)^i (1-0.9)^{3-i}$$

$$= (1 - 0.9)^3 + 3(0.9)(1 - 0.9)^2 + 3(0.9)^2(1 - 0.9) + (0.9)^3$$

$$= 1 .$$

LAPLACE TRANSFORMS

The Laplace transform $F(s)$, of function $f(t)$ is defined as

$$F(s) = \int_0^\infty e^{-st} f(t)\, dt , \tag{2.18}$$

where s is the Laplace transform variable and $f(t)$ is a function. Laplace transforms of selective functions are presented in Table 2.1.

EXAMPLE 2.10

A human operator performs a certain time-continuous task. The operator error times are described by the following probability density function:

$$f(t) = \alpha e^{-\alpha t} \, , \tag{2.19}$$

where t is time and α is the constant error rate. Find the Laplace transform of the above function.

Substituting Eq. (2.19) into (2.18) yields

$$F(s) = \int_0^\infty e^{-st} (\alpha e^{-\alpha t}) \, dt$$

$$= \alpha \int_0^\infty e^{-(s+\alpha)t} \, dt$$

$$= \alpha \left[-\frac{e^{-(s+\alpha)t}}{(s+\alpha)} \right]_0^\infty$$

$$= \frac{\alpha}{s+\alpha} \, . \tag{2.20}$$

EXAMPLE 2.11

Find the Laplace transform of the following function:

$$f(t) = 1 \, . \tag{2.21}$$

With the aid of Eqs. (2.18) and (2.21) we get

$$F(s) = \int_0^\infty e^{-st} [1] \, dt$$

$$= \left[\frac{e^{-st}}{-s} \right]_0^\infty$$

$$= \frac{1}{s} \, . \tag{2.22}$$

QUADRATIC EQUATIONS

Babylonians were solving quadratic equations by 2000 BC. By 1150 AD in India Bhaskara solved quadratic equations by the familiar method of completing the square [1].

Table 2.1. Laplace transforms of selected functions

$f(t)$	$F(s)$
$e^{-\lambda t}$	$\dfrac{1}{s+\lambda}$
$\dfrac{t^{m-1}}{(m-1)!}$, $\quad m = 1, 2, 3, 4, \ldots$	$\dfrac{1}{s^m}$
$\dfrac{df(t)}{dt}$	$sF(s) - f(0))$
$\dfrac{t^{m-1}e^{-\lambda t}}{(m-1)!}$	$\dfrac{1}{(s+\lambda)^m}$, $\quad m = 1, 2, 3, \ldots$
$\dfrac{e^{-\beta t} - e^{-\alpha t}}{\beta - \alpha}$	$\dfrac{1}{(s+\alpha)(s+\beta)}$, $\quad \beta \neq \alpha$
$\displaystyle\sum_{i=1}^{m} \dfrac{G(\lambda_i)}{Q'(\lambda_i)} e^{\lambda_i t}$ The prime denotes the derivative with respect to s	$\dfrac{G(s)}{Q(s)}$, $\quad G(s) = $ polynomial of degree less than m. $Q(s) = (s - \lambda_1)(s - \lambda_2) \ldots (s - \lambda_m)$, where $\lambda_1, \lambda_2, \lambda_3, \ldots \lambda_m$ are all distinct

The quadratic equation is defined below:

$$xs^2 + ys + z = 0 \qquad \text{for } x \neq 0 , \tag{2.23}$$

where x, y and z are constants. The above equation has the following solution:

$$s = \frac{-y \pm (y^2 - 4xz)^{1/2}}{2x} . \tag{2.24}$$

EXAMPLE 2.12

Solve the following equation:

$$s^2 + 9s + 4.25 = 0 . \tag{2.25}$$

In the above equation the values of constants are as follows:

$$x = 1 , \quad y = 9 \quad \text{and} \quad z = 4.25 .$$

Substituting the above values into Eq. (2.24) results in

$$s = \frac{-9 \pm [9^2 - 4(1)(4.25)]^{1/2}}{2(1)} \quad .$$

Therefore,

$$s = -\frac{1}{2}$$

or

$$s = -8.5 \quad .$$

To check the correctness of the above results we substitute them back into Eq. (2.25) as follows:

for $s = -1/2$,

$$\left(-\frac{1}{2}\right)^2 + 9\left(-\frac{1}{2}\right) + 4.25 = 0$$

and for $s = -8.5$,

$$(-8.5)^2 + 9(-8.5) + 4.25 = 0 \quad .$$

The above results prove that the computed values of roots of Eq. (2.25) are correct.

RELIABILITY NETWORKS

This section is concerned with the reliability evaluation of basic configurations such as series, parallel and series–parallel, as shown in Fig. 2.1. In this section it is assumed that each network is composed of independent components or units. The probability-tree technique [8] is used to develop reliability expressions for each of these configurations.

The success or failure of each network element is assigned a probability. The outcome (i.e., success or failure) of each element is represented by the branching limbs of the probability tree. The probability of success for a network is obtained by summing the associated probabilities with the endpoint of the each success path through the probability-tree diagram.

The probability trees for the networks analysed in this section are shown in Figs. 2.2 and 2.3. The number of links in each path from its origin to its

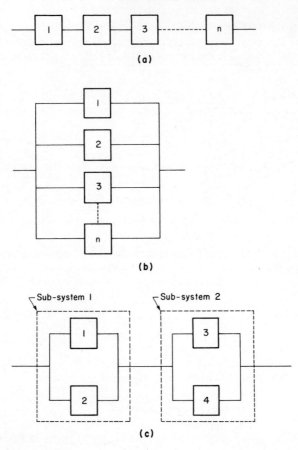

Figure 2.1. Basic reliability network (a) series network; (b) parallel network; (c) series-parallel network.

termination point represent the number of units or devices in the network. The dot on the extreme left side of Figs. 2.2 and 2.3 signifies the point of origin of paths, and the dots on the extreme right side indicates the paths' termination points. The total number of paths in the probability tree is given by 2^n. The n denotes the number of units in a network. The following symbols are used in the probability tree analysis.

S_i denotes the success of the ith unit (i.e., path link), $i = 1, 2, 3, \ldots, n$.

F_i denotes the failure of the ith unit (i.e., path link), $i = 1, 2, 3, \ldots, n$.

R_i denotes the probability of success of the ith unit (i.e., path link), $i = 1, 2, 3, \ldots, n$.

Q_i denotes the probability of failure of the ith unit (i.e., path link), $i = 1, 2, 3, \ldots, n$.

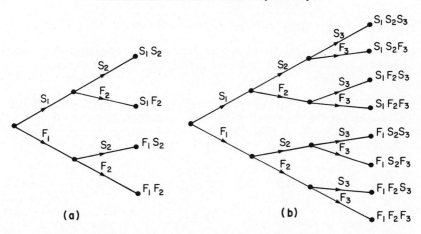

Figure 2.2. Probability tree diagrams (a) for two-nonidentical-unit network; (b) for three-nonidentical-unit network.

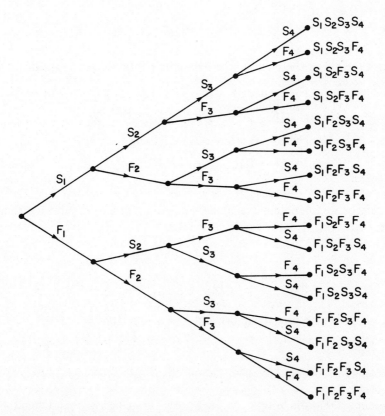

Figure 2.3. Probability tree diagram for a four-nonidentical-unit network.

Parallel Configuration

This network represents a system composed of n independent and active units in parallel, as shown in Fig. 2.1(b). At least one unit must work normally for system success.

The probability tree diagram for a two-nonidentical-unit system is shown in Fig. 2.2(a). From this figure the two-nonidentical-unit parallel-system success paths are $S_1 S_2$, $S_1 F_2$, $F_1 S_2$. With the aid of these paths, the reliability of the two-nonidentical-unit parallel system is given by

$$R_p = R_1 R_2 + R_1 Q_2 + Q_1 R_2 = 1 - (1 - R_1)(1 - R_2) \ , \qquad (2.26)$$

where

$$Q_1 = 1 - R_1 \qquad \text{and} \qquad Q_2 = 1 - R_2 \ .$$

Similarly, for a three-unit parallel system the success paths from Fig. 2.2(b) are

$$S_1 S_2 S_3 \ , \quad S_1 S_2 F_3 \ , \quad S_1 F_2 S_3 \ , \quad S_1 F_2 F_3 \ , \quad F_1 S_2 F_3 \ , \quad F_1 F_2 S_3 \ , \quad F_1 S_2 S_3 \ .$$

With the aid of these paths, the reliability of the three-nonidentical-unit parallel system is given by

$$\begin{aligned}
R_p &= R_1 R_2 R_3 + R_1 R_2 Q_3 + R_1 R_3 Q_2 + R_1 Q_2 Q_3 \\
&\quad + Q_1 R_2 R_3 + Q_1 R_2 Q_3 + Q_1 Q_2 R_3 \\
&= 1 - (1 - R_1)(1 - R_2)(1 - R_3) \ , \qquad (2.27)
\end{aligned}$$

where $Q_i = 1 - R_i$ for $i = 1, 2, 3$.

Similarly, we can obtain a reliability expression for a four-unit parallel network from Fig. 2.3. Thus the generalization of Eq. (2.27) for an n-nonidentical-unit parallel system is given by

$$R_p = 1 - \prod_{i=1}^{n} (1 - R_i) \ , \qquad (2.28)$$

where R_i is the ith unit reliability and $R_i = 1 - Q_i$ for $i = 1, 2, 3, \ldots, n$.

Example 2.13

A system has two independent and active electric motors in parallel. At least one of the motors must work normally for system success. The failure probability of each motor is 0.06. Calculate the parallel system unreliability.

This example indicates that the system is composed of two identical motors. Thus Eq. (2.28) simplifies to

$$R_p = 1 - (1 - R)^2 \ . \tag{2.29}$$

The reliability of each unit is

$$R = 1 - Q = 1 - 0.06 = 0.94 \ .$$

Substituting the above value in (2.29) yields

$$R_p = 1 - (1 - 0.94)^2$$
$$= 0.9964 \ .$$

The unreliability of the parallel system is obtained by subtracting the above result from unity as follows:

$$Q_p = 1 - R_p = 1 - 0.9964$$
$$= 0.0036 \ .$$

Thus the reliability of the parallel system is 0.36%. This result and the end result of Example 2.4 indicate that the system and the task success probabilities have increased in both cases. This is due to the fact that the basic problem of both examples is the same.

Series Configuration

This network represents a system composed of n independent units in series, as shown in Fig. 2.1(a). All of the units must work normally for system success.

From Fig. 2.2(a) the two-nonidentical-unit series-system success path is $S_1 S_2$. Thus, the reliability of the two-nonidentical-unit series system is

$$R_S = R_1 R_2 \ . \tag{2.30}$$

Similarly, from Fig. 2.2(b) the three-nonidentical-unit series-system success path is $S_1 S_2 S_3$. Thus, the reliability of the three-unit series network is

$$R_{S(3)} = R_1 R_2 R_3 \ . \tag{2.31}$$

In the same way, we can obtain a reliability expression for a four-unit series network from Fig. 2.3. Thus the generalization of Eq. (2.31) for an n-nonidentical-unit series system is given by

$$R_s = \prod_{i=1}^{n} R_i \ , \qquad (2.32)$$

where R_i is the ith unit reliability.

EXAMPLE 2.14

An aircraft has two independent and identical engines. The aircraft will crash if any one of the engines fails. Calculate the reliability of both engines working normally, if the reliability of each engine is 0.91. For identical units, Eq. (2.32) becomes

$$R_s = R^n \ . \qquad (2.33)$$

Utilizing the specified data in Eq. (2.33), we get

$$R_s = R^2 = (0.91)^2$$
$$= 0.8281 \ .$$

Thus the reliability of both engines working normally is 82.81%.

Series–Parallel Configuration

The block diagram of this network is shown in Fig. 2.1(c). It is assumed that the network is composed of independent units. In Fig. 2.1(c) the system will fail if either subsystem 1 or subsystem 2 fails, or both subsystems fail.

From Fig. 2.3 the four-nonidentical-unit series–parallel-system success paths are as follows:

$$S_1 S_2 S_3 S_4 \ , \quad S_1 S_2 S_3 F_4 \ , \quad S_1 S_2 F_3 S_4 \ , \quad S_1 F_2 S_3 S_4 \ , \quad S_1 F_2 S_3 F_4 \ ,$$

$$S_1 F_2 F_3 S_4 \ , \quad F_1 S_2 F_3 S_4 \ , \quad F_1 S_2 S_3 F_4 \ , \quad F_1 S_2 S_3 S_4 \ .$$

With the aid of the above paths, the reliability of the four-nonidentical-unit series–parallel system is given by

$$R_{sp} = R_1 R_2 R_3 R_4 + R_1 R_2 R_3 Q_4 + R_1 R_2 Q_3 R_4 + R_1 Q_2 R_3 R_4 + R_1 Q_2 R_3 Q_4$$

$$+ R_1 Q_2 Q_3 R_4 + Q_1 R_2 Q_3 R_4 + Q_1 R_2 R_3 Q_4 + Q_1 R_2 R_3 R_4$$

$$= R_1 R_3 + R_1 R_4 + R_2 R_3 + R_2 R_4$$

$$- (R_1 R_2 R_3 + R_1 R_2 R_4 + R_1 R_3 R_4 + R_2 R_3 R_4) + R_1 R_2 R_3 R_4 \ , \qquad (2.34)$$

where $R_i = 1 - Q_i$ for $i = 1, 2, 3, 4$.

Similarly, for n unit m identical subsystems, the series–parallel network reliability is given by

$$R_{sp} = [1 - (1 - R)^n]^m , \qquad (2.35)$$

where R is the unit reliability.

EXAMPLE 2.15

In Eq. (2.35) the specified values for n, m and R are 2, 2 and 0.95, respectively. Calculate the series–parallel-network reliability.

From Eq. (2.35) we get

$$R_{sp} = [1 - (1 - 0.95)^2]^2$$
$$= 0.9950 .$$

Thus the series–parallel-network reliability is 99.5%.

FINAL-VALUE THEOREM AND MEAN TIME TO FAILURE

Final-Value Theorem

This is used in situations where one is concerned with determining the steady-state value of a function. This theorem is frequently used in the analysis of repairable systems. The Laplace transform of the final-value theorem is

$$\lim_{s \to 0} [sf(s)] , \qquad (2.36)$$

where s is the Laplace transform variable and $f(s)$ is the Laplace transform of function $f(t)$.

EXAMPLE 2.16

The Laplace transform of a function is given by

$$f(s) = \frac{\lambda}{s(s + \lambda + r)} , \qquad (2.37)$$

where λ and r are constants. Find the steady-state value of the above function with the aid of theorem (2.36).

With the aid of theorem (2.36) we get

$$\lim_{s \to 0} \left[s \cdot \frac{\lambda}{s(s + \lambda + r)} \right] = \frac{\lambda}{\lambda + r} . \qquad (2.38)$$

The steady-state value of function (2.37) is $\lambda/(\lambda + r)$.

Mean Time to Failure

This is defined by

$$\text{MTTF} = \int_0^\infty R_s(t)\, dt \ , \tag{2.39}$$

where MTTF is the mean time to failure, $R_s(t)$ is the system reliability function, and t is time.

Alternatively, the mean time to failure of a system is given by

$$\text{MTTF} = \lim_{s \to 0} R(s) \ , \tag{2.40}$$

where $R(s)$ is the Laplace transform of the system reliability.

EXAMPLE 2.17

A system has three independent, identical and active electric motors in parallel. At least one of the motors must work normally for system success. The failure times of a motor are defined by the following probability density function:

$$f(t) = \lambda e^{-\lambda t} \ , \tag{2.41}$$

where λ is the constant failure rate of a motor and t is time.

Assume that the motor failure rate $\lambda = 0.04$ failure/hr. Calculate the system reliability for a 10 hr operation and the system mean time to failure.

From Eq. (2.28), the three-unit parallel-system reliability is given by

$$R_p = 1 - (1 - R)^3 \ , \tag{2.42}$$

where R is the constant motor reliability. By substituting relationship (2.41) into relationship (2.11) the time-dependent motor reliability function is found as

$$R(t) = 1 - \int_0^t \lambda e^{-\lambda t}\, dt$$

$$= e^{-\lambda t} \ . \tag{2.43}$$

After substituting the above equation into (2.42) the system reliability becomes

$$R_p(t) = 1 - (1 - e^{-\lambda t})^3 \ . \tag{2.44}$$

For $\lambda = 0.04$ failures/hr and $t = 10$ hr, the parallel-system reliability is

$$R_p(100) = 1 - [1 - e^{-(0.04)(10)}]^3$$
$$= 0.9642 \ .$$

The three-unit system reliability is 96.42%.

Substituting Eq. (2.44) into Eq. (2.39) and integrating, the following MTTF results:

$$\text{MTTF} = \int_0^\infty [1 - (1 - e^{-\lambda t})^3] \, dt$$
$$= \left[\frac{3e^{-\lambda t}}{-\lambda} + \frac{3}{2\lambda} e^{-2\lambda t} - \frac{e^{-3\lambda t}}{3\lambda} \right]_0^\infty$$
$$= \frac{11}{6\lambda} \ . \tag{2.45}$$

For $\lambda = 0.04$ failures/hr, the above equation yields

$$\text{MTTF} = \frac{11}{6\lambda} = \frac{11}{6(0.04)} = 45.83 \text{ hr} \ .$$

The system mean time to failure is 45.83 hr. If there were no redundancy, the system MTTF would have been 25 hr. Because of redundancy, we were able to increase the MTTF from 25 to 45.83 hr.

SUMMARY

This chapter briefly presents relevant mathematics and basic reliability concepts. Histories of mathematics, probability and reliability engineering are discussed. Some of the basic concepts of probability theory are briefly reviewed, along with exponential and binomial distributions.

The Laplace transform definition is presented. Laplace transforms of selective functions are given in a tabular form. A formula to find the roots of a quadratic equation is given.

Three types of reliability networks are discussed. These are series, parallel and series–parallel networks. The probability-tree technique is used to evaluate the reliability of these three networks. The final-value theorem is presented along with a formula to obtain system mean time to failure.

EXERCISES

1. A person performs 20 tasks/wk. All of the tasks performed by the person during the 1-month period were inspected by an inspector. The inspector found that 10% of the total tasks were performed incorrectly. Calculate the probability of rejecting a task being inspected.
2. Describe the following terms:
   a. independent events
   b. conditional probability
   c. mutually exclusive events.
3. Prove that the probability of the union of two independent events A and B is given by

$$P(A + B) = P(A) + P(B) - P(A)P(B) . \qquad (2.46)$$

   Prove with the aid of a practical example.
4. For the Weibull distribution prove that

$$R(t) + F(t) = 1 , \qquad (2.47)$$

   where $R(t)$ is the reliability function associated with the Weibull distribution and $F(t)$ is the cumulative distribution function associated with the Weibull distribution.
5. Prove that the Laplace transform of $df(t)/dt$ is given by

$$sF(s) - f(0) . \qquad (2.48)$$

6. Find the Laplace transform of the following function:

$$f(t) = 1 - (1 - e^{-\lambda t})^3 , \qquad (2.49)$$

   where λ is a constant.
7. With the aid of Eqs. (2.40) and (2.44) prove that the system mean time to failure is given by

$$\text{MTTF} = \frac{1}{\lambda} \left(1 + \frac{1}{2} + \frac{1}{3} \right) . \qquad (2.50)$$

8. A series–parallel network is shown in Fig. 2.1(c). All of the network units are identical. The constant failure rate of each unit is 0.006 failures/hr. Calculate the network mean time to failure.

REFERENCES

1. H. Eves, *An Introduction to the History of Mathematics.* Holt, Rinehart and Winston, New York (1976).
2. D. E. Smith, *A Source Book in Mathematics.* McGraw-Hill, New York (1929).
3. S. A. Smith, Service reliability measured by probabilities of outage. *Electrical World* **103**, 371-374 (1934).
4. S. A. Smith, Probability theory and spare equipment. *Edison Electric Inst. Bull.* **10**, 85-89 (1934).
5. W. J. Lyman, Fundamental consideration in preparing a master system plan. *Electrical World* **101**, 778-792 (1933).
6. S. A. Smith, Spare capacity fixed for probabilities of outage. *Electrical World* **103**, 222-225 (1934).
7. A. H. S. Ang and W. H. Tang, *Probability Concepts in Engineering Planning and Design.* John Wiley & Sons, New York (1975).
8. B. S. Dhillon and S. N. Rayapati, Reliability evaluation of multi-state device networks with probability trees, in *Proceedings of the Sixth Symposium on Reliability in Electronics*, Hungarian Academy of Sciences, Budapest, Hungary (August 1985).

Chapter 3

Introduction to Human Reliability

INTRODUCTION

Humans play an important role in the overall reliability of engineering systems because various systems are interconnected by human links. This fact was realized by Williams in the later part of the 1950s [1]. He pointed out that in order to have meaningful system reliability analysis, the reliability of the human element must be taken into consideration.

The reliability of the human aspect can be improved significantly by following human-factor principles during the system design phase. On the other hand, factors such as careful selection and training of concerned personnel also help to increase human reliability.

In recent years increasing attention has been paid to several different aspects of the human reliability field, namely data collection and analysis, development of reliability prediction methods and techniques, and so on. An extensive list of references may be found in Ref. [2]. This chapter discusses various areas concerned with the human reliability subject.

STRESS

This is an important area that affects human performance and its reliability. Obviously an overstressed person will have a higher probability of making human errors. According to various researchers [3,4], the relationship between human performance effectiveness and anxiety or stress can be described as shown in Fig. 3.1. The curve shown in Fig. 3.1 indicates that stress is not an entirely negative state. In fact, stress at a moderate level is useful in increasing human effectiveness to its optimal level. Otherwise, at

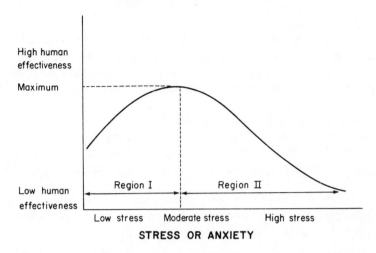

Figure 3.1. Human performance effectiveness versus stress or anxiety.

very low stress, the task will be unchallenging and dull and in turn human performance will not be at its peak. On the other hand, stress above a moderate level will cause human performance to decline. There are various reasons for the decline; for example, worry, fear or other kinds of psychological stress. The moderate stress may be defined as the level of stress enough to keep the human being alert.

The curve shown in Fig. 3.1 is divided into two regions, i.e., regions I and II. In region I, human effectiveness increases with increasing stress, whereas in region II human performance decreases as the level of stress is increased. From Fig. 3.1 it can be concluded that when an operator is performing a task under very high stress, the probability of occurrence of human error will be higher than when he or she is operating under moderate stress.

Occupational Stressors

This section presents the types of occupational stressors. Occupational stressors can be classified into four types as follows [4].

Type I. This is concerned with problems of work load, i.e., work overload or work underload. In the case of work overload the job requirements exceed the individual's ability to meet the requirements. Similarly, in the case of work underload the work performed by one individual fails to provide meaningful stimulation. Examples of work underload are (a) lack of any intellectual input, (b) lack of opportunity to use individual's acquired expertise and skills, and (c) repetitive performance.

Type II. This is concerned with occupational change. This change disrupts the individual's behavioural, physiological and cognitive patterns of functioning. This type of stress is present in an organization concerned with productivity and growth. Some of the forms of occupational change are organizational restructuring, promotion, scientific developments and relocation.

Type III. This is concerned with problems of occupational frustration. This results in situations where the job inhibits the meeting of set goals. Factors such as lack of communication, role ambiguity, bureaucracy difficulties and poor career development guidance form elements of occupational frustration.

Type IV. This is concerned with other possible sources of occupational stressors apart from the above three types; for example, noise, too little or too much lighting and poor interpersonal relationships.

Stress Characteristics of the Human Operator

The human operator has certain limitations in performing a specific task. When these limitations are exceeded, the probability of occurrence of errors rises [5]. In order to minimize the occurrence of human errors operator limitations or characteristics must be considered during the design phase by the design engineer and the reliability engineer. Some operator stress characteristics are given below.

1. Information feedback to the operator is inadequate for the determination of correctness of his or her actions.
2. The operator is required to make comparisons of two or more displays quickly.
3. The operator decision-making time is very short.
4. There is a requirement for prolonged monitoring by the operator.
5. To perform a task, the sequence of steps needed is very long.
6. More than one display are cumbersome to discriminate.
7. There is a requirement to operate more than one control simultaneously at high speed.
8. There is a requirement to perform operator steps at high speed.
9. There is a requirement that decisions have to be made on the basis of data collected from various sources.

Stress Effects on an Individual's Cardiovascular, Muscular and Digestive Systems

Stress has various effects on an individual's body, according to several researchers [4]. Some of these effects are tension headaches, backache, increase in blood pressure, burning sensation in the throat, ulcers, decrease in appetite, tense muscles, coronary heart disease, elevated heart rate and a burning sensation in chest. In addition, diseases and conditions such as strokes, hay fever, fatigue and lethargy, and lowered resistance are stress related.

Stress Checklist Factors

The factors presented in this section function in an additive fashion to increase the stress on the involved person. In fact, these factors are the real problems encountered in the life of a person [4]. Some of these factors are as follows.

1. Having to work with people who have unpredictable temperaments.
2. Unhappy with the present job or work.
3. Serious financial difficulties.
4. Having problems with spouse or children or both.
5. Possibility of redundancy at work.
6. Poor chances for promotion at work.
7. Lacking the expertise to perform the ongoing job.
8. Poor health.
9. Performing under extremely tight time pressures.
10. Taking work home most of the time in order to meet deadlines.
11. Excessive demands from superiors at work.
12. Having a job below one's ability and experience.

ROOK'S MODEL OF HUMAN ERROR OCCURRENCE

This mathematical model of error occurrence was formulated by Rook [6]. The model can be used to compute the total probability of no function failures over all z independent types of tasks. The following assumptions are associated with the model.

1. A number of different tasks are performed to accomplish a mission function.
2. In the mission function, each task may be carried out more than once. In addition, one or more error modes may be associated with a task.
3. The error modes are independent.
4. The entire mission function may or may not fail totally due to an error.

From Ref. [6] the occurrence probability of function failure resulting from the kth error mode of the ith operational task is given by

$$F_{ki} = q_{ki}Q_{ki} \, , \tag{3.1}$$

where q_{ki} is the probability that the ith task arises in an error of the kth mode, and Q_{ki} is the conditional probability that if the mode k error of the *i*th operational type occurs it will result in total function failure. Similarly, from Eq. (3.1), the probability R_{nf} of no function failure is given by

$$R_{nf} = (1 - F_{ki}) \; . \tag{3.2}$$

With the aid of the above equation, the no-function-failure probability, R_{nfi}, resulting from one task of the *i*th type over all the M_i different error modes is given by

$$R_{nfi} = \prod_{k=1}^{M_i} (1 - q_{ki} \cdot Q_{ki}) \; . \tag{3.3}$$

The probability of no failures, R_{nfti}, over all s_i *i*th type-independent tasks, with the aid of Eq. (3.3), is

$$R_{nfti} = (R_{nfi})^{s_i} \; , \tag{3.4}$$

where s_i is the number of times the *i*th type of task is performed in the total function.

The resulting overall probability, R_T, of no function failures over all z independent types of tasks, is

$$R_T = \prod_{i=1}^{z} R_{nfti} \; , \tag{3.5}$$

where z are the distinct types of tasks in the function. Substituting Eq. (3.4) into Eq. (3.5) results in

$$R_T = \prod_{i=1}^{z} (R_{nfi})^{s_i} \; . \tag{3.6}$$

This mathematical model fits quite well in situations where system failures result from errors committed in repetitive manual assembly work. An example of such errors is the "omitted solder joint."

HUMAN PERFORMANCE RELIABILITY MODELING

This section is concerned with human performance reliability modeling in a continuous time domain [7–9]. This modeling concept is analogous to classical reliability modeling. Thus the general human performance reliability function for time-continuous human tasks can be developed the same way as for the case of the classical reliability function. Examples of time-continuous tasks performed by humans are aircraft maneuvering, missile countdown and scope monitoring.

Generalized Human Performance Reliability Function

The time-dependent human error (hazard) rate, $h_{e(t)}$, is defined as follows [10]:

$$h_e(t) = - \frac{1}{R_e(t)} \frac{dR_e(t)}{dt} , \tag{3.7}$$

where $R_e(t)$ is the human performance reliability at time t. Rearranging Eq. (3.7) results in

$$h_e(t) \, dt = - \frac{1}{R_e(t)} \, dR_e(t) . \tag{3.8}$$

By integrating both sides of the above equation over the time interval $[0, t]$ we obtain

$$\int_0^t h_e(t) \, dt = - \int_{R_e(0)}^{R_e(t)} \frac{1}{R_e(t)} \, dR_e(t) , \tag{3.9}$$

where $R_e(0) = 1$ at $t = 0$. After evaluating the right-hand side of (3.9), the resulting expression becomes

$$\ell n \, R_e(t) = - \int_0^t h_e(t) \, dt ,$$

$$R_e(t) = \exp \left[- \int_0^t h_e(t) \, dt \right] . \tag{3.10}$$

The above reliability expression holds whether the human error rate is constant or nonconstant. In other words, it holds when the human error (hazard) rate is described by statistical distributions such as the exponential, gamma, Rayleigh, Weibull, normal or bathtub distributions [11]. The data for the time-continuous tasks were collected experimentally in Ref. [9]. These data were obtained from an experiment concerning a vigilance task. In this experiment, the subjects were asked to observe a clock-type display of lights. In addition, when a failed light event occurred the subject responded to it by pressing a hand-held switch. One example of data collected from this study is miss error. The miss error is associated with a situation when the subject failed to detect the failed light. The Weibull function was one of the three distributions which fitted quite well to the experimental data of Ref. [9].

The general expression to obtain the mean time to human error (MTHE) is

$$\text{MTHE} = \int_0^\infty \exp\left[-\int_0^t h_e(t)\, dt\right] dt \ . \tag{3.11}$$

EXAMPLE 3.1

The constant human-error rate, h_e, is associated with a time-continuous task. Obtain an expression for the human-performance reliability (i) with the aid of Eq. (3.10); (ii) with the aid of the Markov method.
In this example the human error (hazard) rate, $h_e(t)$, is

$$h_e(t) = h_e \ . \tag{3.12}$$

By substituting Eq. (3.12) into Eq. (3.10) we get

$$R(t) = \exp\left[-\int_0^t h_e\, dt\right]$$
$$= \exp[-h_e t] \ . \tag{3.13}$$

In the second case, the human performance reliability expression is developed with the aid of Fig. 3.2. The Markov approach is described in detail in Ref. [10].
The assumptions associated with the Markov technique are as follows.

1. All of the occurrences are independent.
2. The probability of occurrence of a transition, say, from state α to state $(\alpha + 1)$ in the finite time interval Δt, is h_e times Δt (i.e., $h_e \Delta t$), where h_e is the transition rate. In Fig. 3.2 h_e represents a constant human-error rate.
3. In a time interval Δt the probability of more than one occurrence is negligible.

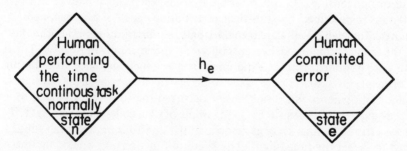

Figure 3.2. State-space diagram.

The following symbols are associated with the model: $P_n(t)$ is the probability that the human is performing the time-continuous task normally at time t, $P_e(t)$ is the probability that the human has committed an error at time t, and s is the Laplace transform variable.

With the aid of the Markov approach, we write down the following equations for Fig. 3.2:

$$P_n(t + \Delta t) = P_n(t)(1 - h_e \Delta t) \ , \tag{3.14}$$

$$P_e(t + \Delta t) = P_e(t) + P_n(t) h_e \Delta t \ . \tag{3.15}$$

Rearranging Eqs. (3.14) and (3.15) and taking the limit as $\Delta t \to 0$ we get

$$\lim_{t \to 0} \frac{P_n(t + \Delta t) - P_n(t)}{\Delta t} = \frac{dP_n(t)}{dt} = -h_e P_n(t) \tag{3.16}$$

and

$$\lim_{t \to 0} \frac{P_e(t + \Delta t) - P_e(t)}{\Delta t} = \frac{dP_e(t)}{dt} = P_n(t) h_e \ . \tag{3.17}$$

At $t = 0$, $P_n(0) = 1$ and $P_e(0) = 0$.

Solving Eqs. (3.16) and (3.17) with the aid of Laplace transforms, we get

$$P_n(t) = \frac{1}{s + h_e} \tag{3.18}$$

and

$$P_e(t) = \frac{h_e}{s(s + h_e)} \ . \tag{3.19}$$

After taking the inverse Laplace transform of the above equations we get

$$P_n(t) = e^{-h_e t} \tag{3.20}$$

and

$$P_e(t) = (1 - e^{-h_e t}) \ . \tag{3.21}$$

Thus from Eq. (3.20), the human performance reliability, $R_e(t)$, at time t is

$$R_e(t) = P_n(t) = e^{-h_c t} \; . \tag{3.22}$$

The above equation is the same as Eq. (3.13). It means that both methods are correct.

EXAMPLE 3.2

The time-dependent human error rate, $h_e(t) = \theta t^{\theta-1}/\beta^\theta$, is associated with a time-continuous task. Obtain an expression for the human performance reliability. The symbols β and θ are known as the scale and shape parameters, respectively.

In this example the time-dependent human error rate expressions indicate that the error (hazard) rate is associated with the Weibull distribution. In any case, substituting the error rate expression into Eq. (3.10) results in

$$R_e(t) = \exp\left[-\int_0^t \frac{\theta t^{\theta-1}}{\beta^\theta} \, dt \right]$$

$$= \exp\left[-\left(\frac{t}{\beta}\right)^\theta \right] \; . \tag{3.23}$$

Thus Eq. (3.23) is the human performance reliability function.

Human Performance Correctability Function for Time-Continuous Tasks

This function is concerned with the human capacity to correct self-generated human errors. According to Ref. [8] the correctability function may be defined as the probability that an error will be corrected in time t subject to stress constraint associated with the task and its environment. Mathematically, the correctability function is defined as below:

$$P_c(t) = 1 - \exp\left[-\int_0^t r_c(t) \, dt \right] \; , \tag{3.24}$$

where $P_c(t)$ is the probability that an error will be corrected in time t and $r_c(t)$ is the time-dependent rate at which tasks are corrected.

The above correctability function holds whether the task correction rate is constant or nonconstant. More clearly, it holds whether the correction rate is described by the exponential distribution or any other statistical distributions of a continuous random variable. Some of the examples of other distributions are Rayleigh, Weibull and normal statistical functions.

Alternatively, the correctability function can be expressed as follows:

$$P_c(t) = \int_0^t f(t)\, dt \ , \tag{3.25}$$

where $f(t)$ is the probability density function associated with the time-to-correction completion.

EXAMPLE 3.3

The following probability density function is associated with the time-to-correction completion:

$$f(t) = \frac{\theta}{\beta^\theta} t^{\theta-1} \exp\left[-\left(\frac{t}{\beta}\right)^\theta\right] \quad \text{for } \theta > 0 \ , \quad \beta > 0 \ , \quad t \geq 0 \ , \tag{3.26}$$

where t is time, β is the scale parameter, and θ is the shape parameter. Obtain an expression for the correctability function.

By substituting Eq. (3.26) into Eq. (3.25) we get

$$P_c(t) = \int_0^t \frac{\theta}{\beta^\theta} t^{\theta-1} \exp\left[-\left(\frac{t}{\beta}\right)^\theta\right] dt \ . \tag{3.27}$$

The following expression results from Eq. (3.27):

$$P_c(t) = 1 - \exp[-(t/\beta)^\theta] \ . \tag{3.28}$$

One should note here that the above human-performance correctability function is associated with the Weibull distribution.

FAULT TREE METHOD

This is a widely used method to perform reliability analysis of engineering systems. This method was originally developed by H. W. Watson of Bell Laboratories in 1961 to perform reliability and safety analysis of the Minuteman launch-control system. Since those early days various other researchers have refined this method. The method is quite useful for analyzing complex engineering systems. In recent years it has been frequently applied to analyze nuclear power generation systems. The method is described in detail in Refs. [10,12]. A list of extensive references is given in Ref. [13].

The starting point of the fault tree analysis is the identification of the undesirable event of the system. This event is also known as the top event. The top event is caused by events generated and connected by logic gates. Two examples of logic gates are AND and OR gates. Basic steps used to perform fault tree analysis are shown in Fig. 3.3.

The following symbols are associated with a basic fault tree.

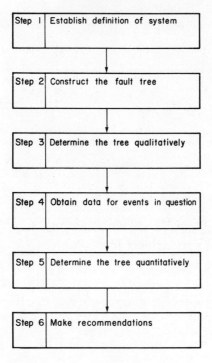

Figure 3.3. Fault tree analysis steps.

1. *Circle.* This represents basic fault events; more specifically, those events that need not to be developed any further.
2. *Diamond.* This represents those fault events that will not be developed further because of lack of interest or data.
3. *Rectangle.* This represents a fault event that results from the combination of fault events through the input of a logic gate, such as AND or OR gates.
4. *AND gate.* The symbol for this logic gate is shown in Fig. 3.4. This gate signifies that an output fault event occurs if all of its (i.e., gate) input fault events occur.
5. *OR gate.* The symbol for this logic gate is shown also in Fig. 3.4. This gate signifies that an output fault event occurs if one or more of the input fault events occur.

EXAMPLE 3.4

A person is required to perform a task, say, X. This task is composed of three independent subtasks, A, B and C. All of these subtasks must be performed correctly for overall task success. Subtask A is composed of three steps, a_1, a_2 and a_3. Each of these steps must be performed correctly irrespective of

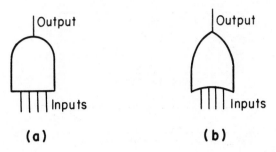

Figure 3.4. Basic logic gate symbols (a) AND gate, (b) OR gate.

their sequence. Subtask B is composed of only two steps, i.e., b_1 and b_2. This subtask will be performed correctly if at least one of the b_1 and b_2 steps is performed correctly. In addition, each of these two steps b_1 and b_2 is composed of two substeps. Both substeps associated with a step must be accomplished successfully for the success of the step. The substeps β_1, β_2 and α_1, α_2 are associated with steps b_1 and b_2, respectively. If subtasks, steps and substeps are independent, develop a fault tree with the top event entitled "The person will not perform task X successfully." The fault tree is shown in Fig. 3.5.

Probability Evaluation of a Fault Tree

This section deals with the quantitative evaluation of a fault tree. The probability of the occurrence of the output fault event of AND and OR gates can be calculated from the following two equations.

$$\text{AND gate:} \qquad F_0 = f_1 f_2 f_3 \ldots f_n \,, \qquad (3.29)$$

where F_0 is the probability of the occurrence of the output fault event; f_i is the probability of the occurrence of the independent input fault event i for $i = 1, 2, 3, \ldots n$; and n is the number of input fault events.

$$\text{OR gate:} \qquad F_{00} = 1 - (1 - f_1)(1 - f_2) \ldots (1 - f_n) \,, \qquad (3.30)$$

where F_{00} is the probability of the occurrence of the output fault event; f_i is the probability of the occurrence of the independent input fault event i for $i = 1, 2, 3, \ldots n$; and n is the number of input fault events.

Example 3.5

In Fig. 3.5, the probability of occurrence of each basic fault event is assumed to be 0.002. Calculate the probability of occurrence of the top fault event, i.e., the person will not perform the task X successfully.

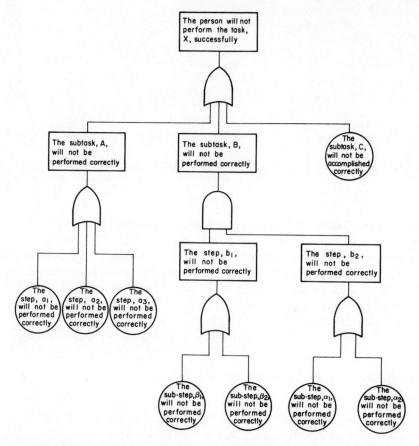

Figure 3.5. A fault tree for the unsuccessful performance of task X.

The fault tree shown in Fig. 3.5 is redrawn in Fig. 3.6 with specified values for basic fault events. In the fault tree diagram the gates are labeled with capital letters A, B, C, D and E.

The probability of occurrence of the output fault event of gate A with the aid of Eq. (3.30) is

$$F_{00} = 1 - (1 - f_1)(1 - f_2)(1 - f_3)$$
$$= 1 - (1 - 0.002)(1 - 0.002)(1 - 0.002)$$
$$= 0.006 .$$

Similarly, the probabilities of occurrence of output fault events of gates B and C are 0.004 and 0.004, respectively.

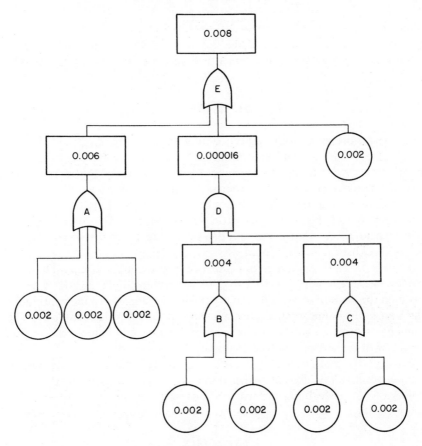

Figure 3.6. Probability tree.

With the aid of Eq. (3.29), the probability of occurrence of the output fault event of gate D is given by

$$F_0 = f_1 f_2 = (0.004)^2$$
$$= 0.000016 \ .$$

Finally, the probability of occurrence of the output fault event of gate E (i.e., the top event of the fault tree), with the aid of Eq. (3.30), is

$$F_{00} = 1 = (1 - f_1)(1 - f_2)(1 - f_3)$$
$$= 1 - (1 - 0.006)(1 - 0.000016)(1 - 0.002)$$
$$= 0.008 \ .$$

Thus the probability of occurrence of the top fault event known as "the person will not perform the task X successfully" is 0.008. In other words, there is only a 0.8% chance that the person will not perform the task X correctly.

SUMMARY

This chapter presents various aspects of human reliability. The topic of stress is discussed. The human performance effectiveness versus stress curve is briefly described with the aid of a diagram. It is concluded that at moderate stress human performance effectiveness is at its maximum level. Four different types of occupational stressors are briefly described along with the stress characteristics of the human operator. Various stress effects on an individual's cardiovascular, muscular and digestive systems are discussed. Twelve stress checklist factors are listed.

Rook's model of human-error occurrence is presented. A general human-performance reliability function for time-continuous human tasks is developed on the basis of classical reliability theory. Examples of time-continuous tasks are scope monitoring, aircraft maneuvering and missile countdown. The human-performance correctability function for time-continuous tasks is presented with the aid of a solved example.

Finally, the fault tree technique is discussed. This method can be used in human unreliability analysis. An example is presented to calculate the probability of not performing the required task successfully.

EXERCISES

1. What are the advantages and disadvantages of the fault-tree method with respect to human reliability analysis?
2. Describe the relationship between human performance effectiveness and stress.
3. What are the causes of occupational stress?
4. Discuss the effects of stress on individual's body.
5. List at least five practical examples of time-continuous human tasks.
6. A person is performing a time-continuous task. The human error rate associated with that task is 0.04 error/hr. Calculate the human reliability for a 4-hr mission.
7. Explain the following four terms associated with the fault-tree method:
   a. top event
   b. logic gate
   c. basic event
   d. fault tree symbols.

REFERENCES

1. H. L. Williams, Reliability evaluation of the human component in man-machine systems. *Electrical Manufacturing* **4**, 78–82 (1958).
2. B. S. Dhillon, On human reliability – bibliography. *Microelectronics and Reliability* **20**, 371–373 (1980).
3. E. W. Hagen (Ed.), Human reliability analysis. *Nuclear Safety* **17**, 315–326 (1976).
4. H. R. Beech, L. E. Burns and B. F. Sheffield, *A Behavioural Approach to the Management of Stress*. John Wiley & Sons, Chichester (1982).
5. D. Meister, Human factors in reliability, in *Reliability Handbook* (Edited by W. G. Ireson, pp. 12.2–12.37. McGraw-Hill, New York (1966).
6. L. W. Rook, *Reduction of Human Error in Industrial Production*. Report No. SCTM 93-62(14), Sandia Laboratories, Albuquerque, New Mexico (June 1962).
7. W. B. Askren and T. L. Regulinski, Quantifying human performance for reliability analysis of systems. *Human Factors* **11**, 393–396 (1969).
8. T. L. Regulinski and W. B. Askren, Stochastic modeling of human performance effectiveness functions, in *Proceedings of the Annual Reliability and Maintainability Symposium*, IEEE, New York, pp. 407–416 (1972).
9. T. L. Regulinski and W. B. Askren, Mathematical modeling of human performance reliability, in *Proceedings of Annual Symposium on Reliability*, IEEE, New York, pp. 5–11 (1969).
10. B. S. Dhillon and C. Singh, *Engineering Reliability: New Techniques and Applications*. John Wiley & Sons, New York (1981).
11. B. S. Dhillon, Life distributions. *IEEE Transactions on Reliability* **30**, 457–460 (1981).
12. B. S. Dhillon, Fault tree analysis, in *Mechanical Engineer's Handbook* (Edited by Mayer P. Kutz), Chapter 20. John Wiley & Sons, New York (1985).
13. B. S. Dhillon and C. Singh, Bibliography of literature on fault trees. *Microelectronics and Reliability* **17**, 501–503 (1978).

Chapter 4

Human Errors

INTRODUCTION

Humans play an important role during the design, production and operation phases of a system. The degree of their role may vary from one system to another and from one system phase to another. This human role is subject to deterioration as a result of human error. The overall reliability of a system is affected because humans have some probability of performing their normal tasks incorrectly. According to a study quoted by Meister [1] from 20 to 50% of all equipment failures are due to human errors. This is an alarming proportion of total equipment failures. This means that the reliability of the human element has to be included in reliability analysis in order to obtain a correct picture of the system reliability. This fact was long ago recognized by Williams [2]. Ever since the Williams article many more articles have appeared on human reliability [3].

This chapter is concerned with human errors. A number of selective documents associated with human error are listed at the end of this chapter [1–29].

REASONS FOR HUMAN ERRORS AND THEIR CONSEQUENCES

There are various reasons for the occurrence of human errors. According to Rigby [20], "most of the human errors occur because humans are capable of doing so many different things in many diverse ways." However, according to Meister [1] some of the reasons are more specifically as follows:

1. Inadequate lighting in the work area.
2. Inadequate training or skill of the concerned manpower; for example, operators, maintenance and production personnel.
3. Poor equipment design.
4. High temperature in the work area.
5. High noise level.
6. Inadequate work layout.
7. Crowded work space.
8. Poor motivation.
9. Improper tools.
10. Poorly written equipment maintenance and operating procedures.
11. Inadequate handling of equipment.
12. Poor management.
13. Task complexity.
14. Poor verbal communication.

The consequence of human errors may vary from one set of equipment to another or one task to another. Furthermore, a consequence may range from minor to severe (for example, from delay in system performance to loss of life). However, in broad terms the consequence of a human error with respect to equipment may be classified into the following three categories: *Category I:* Equipment operation is prevented; *Category II:* Equipment operation is delayed significantly but not prevented; and *Category III:* Equipment operation delay is insignificant.

OCCURRENCE OF HUMAN ERRORS

Human error occurs in various ways. These are as shown in Fig. 4.1 according to Hammer [18].

CLASSIFICATION OF HUMAN ERRORS

Human errors may be broken down into various categories [1,15]. Many of these categories are listed below:

1. operating errors
2. assembly errors
3. design errors
4. inspection errors
5. installation errors
6. maintenance errors.

Each of the above types is described in the following sections.

HR-C

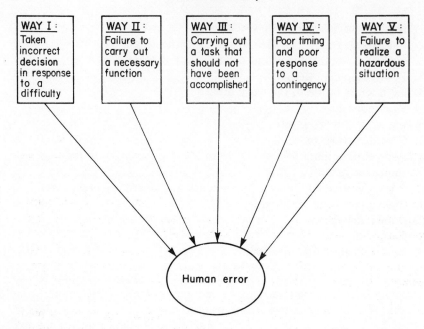

Figure 4.1. Ways in which a human error occurs. (Adapted from Hammer [18].)

Operating Errors

These are concerned with errors due to operating personnel. Almost all of the operating errors occur in the field-use environment. Situations such as those that follow lead to operating errors [15]:

1. Lack of proper procedures.
2. Task complexity and overload conditions.
3. Poor personnel selection and training.
4. Operator carelessness and lack of interest.
5. Poor environmental conditions.
6. Departure from following the correct operating procedures.

Function-associated errors. An operator can carry out various types of functions. These are decision making, sequencing, problem solving, estimating and tracking; and detecting, identifying, sensing, classifying and coding. According to Altman [4], many potential errors are associated with all of these functions. For example, with the decision-making function potential errors such as premature decision making, application of an unnecessary rule and failure to apply an available rule are possible. Potential errors such as premature response to a target change, incorrect control-action direction and late response to a target change are associated with the estimating and tracking functions.

The function of sequencing is prone to potential errors such as insertion of an undesirable step, omission of a procedural step, and so on. A typical example of the potential error associated with the problem-solving function is the formulation of erroneous rules. Finally, the sensing, identifying, detecting, classifying and coding functions are also subject to many potential errors. Some of them are failure to record a deviation in signal, or to monitor and report the appearance of a target.

Human errors in operating equipment. Lincoln [7] has classified the human errors made in operating equipment into two categories, i.e., errors of omission (category I) and errors of commission (category II). The category I errors are composed of errors of attention and errors of memory. Errors of attention are associated with situations requiring operator attention. A typical example of such situations is an operator expected to notice changes in values displayed on a group of meters. Failure to notice a change is regarded as the attention error. Similarly, as the name suggests, the errors of memory are concerned with human memory. When the operator forgets to perform a task it is considered a memory error.

The category II errors consist of errors of operation, errors of identification and errors of interpretation. First, an operation error may be described as one in which the control movement is unnecessary to achieve the desired effect. Second, the error of identification is associated with misidentification of an object and its treatment as the correct object. According to Lincoln [7], the frequency of occurrence of this type of error is higher than that of any of the other types of errors. Finally, errors of interpretation are concerned with misunderstanding of information, and result in performing incorrect tasks.

Assembly Errors

These errors are caused by humans and occur during product assembly. They are the result of poor workmanship. Many times, assembly errors are discovered in the field environment after experiencing a failure. Some of the examples of the assembly errors are as follows:

1. Using an incorrect component.
2. Omitting a component.
3. Assembly incompatible with blueprints.
4. Incorrect soldering.
5. Part is wired backwards.

According to Meister [15], there are many causes of production errors. Most of them are included in the second section (Reasons for Human Errors) of this chapter. Some of the causes are as follows:

1. inadequate illumination
2. excessive noise level
3. poorly designed work layout
4. poor communication of information and excessive temperature
5. inadequate supervision and training
6. poor blueprints.

Design Errors

These types of errors are due to inadequate design by the designers. The three types of errors are the failure to implement human needs in the design, assigning an inappropriate function to a person and failure to ensure the effectiveness of the man and machine component interactions. Factors such as too much hastiness in the design effort, inclination of the designer to a particular design solution and poor analysis of the system needs are the causes of design errors.

Inspection Errors

These errors are associated with inspection. The objective of the inspection is to uncover defects. Inspection errors occur because the inspection is not 100% accurate. Due to an inspection error an in-tolerance component may be rejected or an out-of-tolerance component accepted. According to one study reported by McCornack [21], an average inspection effectiveness is close to 85%.

Installation Errors

These errors occur during the installation stage of equipment. These errors are short-term errors. One of the prime causes of installation errors is the failure to install equipment by humans according to instructions or blueprints.

Maintenance Errors

These errors result in the field due to incorrect repair of the equipment in question. Examples of maintenance errors are incorrect calibration of equipment, application of the wrong grease at appropriate points of the equipment, and so on. According to Meister [15], the occurrence of maintenance errors may increase as the equipment becomes older. This is the result of the increase in maintenance frequency because of wear-out.

HUMAN ERROR PROBABILITY ESTIMATES

Human error probability is the primary measure of human performance. This is defined by Green and Bourne [22] as follows:

$$\text{human error probability} = P_{he} = \frac{E_n}{O_{pe}} \ . \qquad (4.1)$$

where O_{pe} is the total amount of opportunities for the error; E_n is the total amount of known errors of a given type; and P_{he} is the probability that when a specified task is carried out a human error will occur.

From Ref. [23] the estimated values of P_{he} for selective tasks are presented in Table 4.1. Before using these tabulated values for human error probability in real-life situations, one should consult Ref. [23] for associated assumptions.

HUMAN ERROR IN AIR TRAFFIC CONTROL SYSTEM OPERATIONS AND IN NUCLEAR POWER PLANTS

Both of these two areas are prone to human errors, with severe consequences. Therefore, this section briefly examines the occurrence of human errors in these two important industrial areas separately.

Table 4.1. Human error probability estimates for selected tasks

Task No.	Task description	Human error probability
1	Reading a chart recorder	0.006
2	Reading an analog meter	0.003
3	Reading graphs	0.01
4	Interpreting incorrectly the indication on the indicator lamps (checked individually for some specific purpose)	0.001
5	Turning a control in the incorrect direction under high stress	0.5
6	Using the checklist correctly	0.5
7	Mating a connector	0.01
8	Choosing an incorrect panel control out of a number of similar controls	0.003

Air Traffic Control System Errors

The possibility of aircraft collisions in the United States was realized almost four decades ago when the first air traffic control center in New Jersey was established [24]. However, it was not until 1958 that the Federal Aviation Administration (FAA) was established to look after the safe and effective use of United States airspace. This was the result of a Trans World Airlines Super Constellation and United Airlines DC-7 collision over the Grand Canyon in 1956.

According to Ref. [24], in field facilities a work force of about 27,000 air traffic controllers is employed by the FAA. Furthermore, about 14,000 aircraft flights (alone) are handled by the air traffic control system on a typical day. This factor alone makes one realize the criticality of human error in air traffic control system operation. For example, according to Kinney, Spahn and Amato [25] over 90% of the documented air traffic control system errors were due to humans. Specifically, these were directly due to controllers and first-line supervisors. In addition, all of these human errors were associated with (1) communications, (2) judgment, and (3) attention. Air traffic system error [24] is expressed as an operational error in which a failure of the human element, equipment, procedural and/or other system components, in combination or individually, leads to less than the necessary separation minima outlined in the authoritative air traffic control documents.

System-error contributing causes may be categorized as follows:

1. communication
2. stress
3. operations management
4. procedures
5. attention
6. equipment
7. external factors and environment
8. judgment.

However, according to the study of Kinney, Spahn and Amato [25] inappropriate control techniques and work habits are the main factors causing system errors. These factors are the result of the following items:

1. incomplete technical supervision;
2. poor awareness of the usefulness of good work habits;
3. lack of detail description in written standard operating procedures.

Air traffic control system functions and controller tasks. The system is required to perform functions as shown in Fig. 4.2. In all four basic system functions the aircraft controller is an important component. A wide variety of tasks

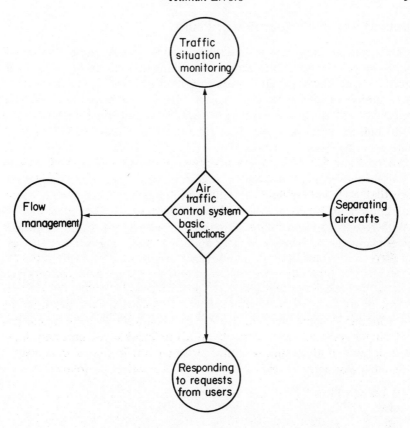

Figure 4.2. Air traffic control system basic functions.

associated with these functions are carried out by the controller. Examples of such tasks are as follows:

1. Communication through radio and interphone.
2. Close coordination with fellow workers.
3. Choose plans and strategies as necessary.
4. Revise plans and strategies to satisfy needs.
5. Perform data entries.
6. Observe aircraft directly.
7. Monitor aircraft blips on radar.
8. Operate controls and read flight progress strips.

Tasks such as those listed above have to be performed effectively by the controllers in order to minimize human errors.

Human Error in Nuclear Power Plants

Humans play an important role in the reliability, availability and safety of nuclear power plants. Events leading to unplanned radioactivity release and forced outages due to operator errors have resulted in the past. Furthermore, on several occasions in nuclear power plants component failures have occurred due to humans. According to Joos, Sabri and Husseiny, [26] a total of 401 human errors occurred from June 1, 1973 to June 30, 1975 in U.S. commercial light-water reactors. The well-publicized accidents (e.g., Three Mile Island and Brown's Ferry) witness to the fact that humans have acted in nuclear power plants as accident initiators, propagators and mitigators. Therefore, one may say that the problem of human error in a nuclear power generation plant is crystal clear and has to be dealt with effectively.

With respect to the availability of engineered safety features in nuclear power plants the following two tasks are of particular importance: (1) control room operations, and (2) maintenance and calibration procedures.

The reason for their importance is their potential impact on the availability of engineered safety features. When performing human reliability analysis in nuclear power stations factors such as those that follow are to be taken into account [27]:

1. psychological stress level
2. training quality
3. quality of method of use
4. human redundancy and human action independance
5. quality of documented instructions
6. display feedback type
7. human engineering quality with respect to displays and controls.

HUMAN ERROR PREVENTION METHODS

This section briefly discusses four methods to reduce and prevent human errors. Strictly speaking, the fourth method presented in the section is not really a method but a number of preventive measures to prevent specific causes of operator errors.

Method I

This method is known as man–machine systems analysis. The original version of it is due to Robert B. Miller. He developed a technique entitled "a method for man–machine task analysis" in the early 1950s [19,28]. The man-machine systems analysis method (MMSAM) is used to lower human-error-caused unwanted effects to some acceptable level in a system. This method is composed of 10 steps as shown in Fig. 4.3. Steps 1 and 4–10 are considered to be self-explanatory. Their description is therefore omitted.

Step 2 is concerned with those performance-shaping factors (i.e., situational characteristics) under which humans will have to carry out various tasks

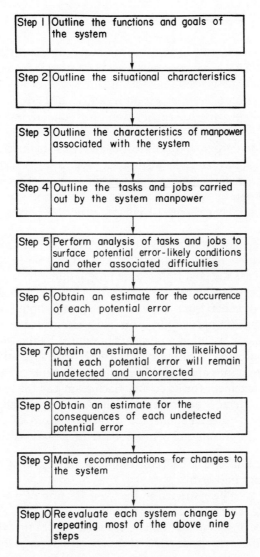

Figure 4.3. Steps of the man–machine system analysis procedure. (Adapted from Miller [28].)

and jobs. Typical examples of such performance-shaping factors are illumination, union actions, quality of the air, general cleanliness, and so on.

Step 3 is concerned with the identification and estimation of the characteristics of the manpower associated with the system. Examples of such characteristics are training, experience, motivation and skills.

Method II

This method is known as the error-cause removal program (ECRP) [16]. The emphasis of the ECRP is on preventive measures rather than merely on remedial ones. This method is useful in reducing human error to some tolerable level once production operations get under way. Furthermore, the method is also useful in improving job satisfaction of production workers because it requires their direct participation. Therefore, it can simply be called the worker-participation program to reduce human errors. In the data collection, analysis, and design recommendation aspects, production workers are directly involved. This direct participation makes production workers see this error-cause removal program as their own. Examples of the production workers are machinists, assembly personnel, handlers, inspection personnel and maintenance workers.

The ECRP is composed of teams of production workers. Each team has a coordinator whose responsibility is to keep the group's activities goal directed. In this case the goal is to reduce production errors. These coordinators possess special technical and group skills but they can be either production workers or supervisors. The size of the group should not be greater than 8 or 12 persons. In the periodic error-cause removal team meetings, the error and error-likely situation reports are presented by the workers. These reports are examined and discussed and then at the end, suggestions for remedial or preventive actions are made. The team coordinator presents the team proposals to management for its action. Each team and the management is assisted by human-factors and other specialists. These specialists help both parties with respect to evaluations and implementations of the suggested design solutions. Important guidelines for the ECRP are as follows:

1. The data collection is to be concerned with error-likely situations, accident-prone situations and errors.
2. The program is to be restricted to the identification of work conditions that need redesign to reduce error potential.
3. A team of specialists should evaluate each redesign of the work situation suggested by the error-cause removal team with respect to factors such as amount of reduction of errors and increase in job satisfaction and cost effectiveness.

Error-cause removal program components. The error-cause removal program consists of the following basic elements:

1. The most appropriate design solutions are implemented by the management.
2. The efforts of production workers in the error-cause removal program is recognized appropriately by the management.
3. Each person involved with the error-cause removal program is educated about its usefulness.
4. The proposed design solutions are evaluated in terms of cost and worth by human factors and other specialists. In addition, they choose the most appropriate of these solutions or develop alternative solutions.
5. The error-cause removal team coordinators and production workers are trained in the techniques to be used for data collection and analysis.
6. The effects of changes in the production process are evaluated by human-factors and other specialists aided by continuing inputs from the ECRP.
7. The errors and error-likely situations are reported and their causes determined by the production workers. In addition, to remove or appropriately modify these error causes, design solutions are proposed by the workers.

Method III

This method is known as quality-control circles. In 1963, the method was developed in Japan to find solutions to quality-control problems. Its applications have been quite successful in Japan.

The quality-control circles and the error-cause removal program methods have much in common and certain of their elements are much the same. Some examples of such elements [14,29] are as follows:

1. participative democracy concept
2. orientation towards problem solving
3. crossover among levels of management.

Some of the ways in which the quality-circle method differs from the ECRP are as follows:

1. Investigation of problems with the aid of cause–effect diagrams and Pareto analysis.
2. Emphasis on teamwork and identification with the company.
3. Emphasis on training in statistical quality-control methods.

A circle on a voluntary basis is formed by 8–10 persons. These persons are production engineers, supervisors and workers carrying out interrelated or similar work. All of these people are provided training in statistical quality-control methods. Some of the areas covered in the training are listed below:

1. cause–effect diagrams
2. quality-control charts
3. Pareto diagrams
4. histograms
5. binomial distribution.

The cause–effect diagram was first developed in 1950 by a Japanese named Ishikawa. This type of diagram is developed by defining an effect and then reducing it to its contributing factors, known as causes. The procedure followed to develop a cause–effect diagram is to first list pertinent causes in terms of the following four classifications:

1. materials
2. people
3. machines
4. techniques.

In addition, the pertinent causes are iteratively reduced to their subcauses. The process is only terminated when all of the possible causes are listed. All of the causes involved are analysed carefully with respect to their contribution to the effect.

The Pareto analysis is based on Pareto's principles. Vilfredo Pareto was an Italian sociologist and economist. He lived from 1848 to 1923. With respect to quality control his principle states that there are always a few kinds of defects in hardware manufacture, which loom large in frequency of occurrence and severity. Pareto's principle helps to identify the area where effort has to be concentrated. This principle can also be applied in human-error analysis.

Information on quality-control charts, the binomial distribution and histograms can be found in standard quality-control textbooks.

Method IV

As mentioned earlier, this is not really a method but a number of preventive measures to prevent causes of operator errors. There are many causes of operator errors. The intent of this section is to discuss preventive measures for selected causes of operator errors. Therefore, preventive measures for the following causes of operator errors [18] are discussed in this section:

1. inattention
2. fatigue
3. failure to note important indications
4. imprecise control setting by operator

5. controls switched on in incorrect sequence
6. mistake in reading instruments
7. mistake in use of controls
8. vibration and noise irritation
9. failure to act at the moment of need due to faulty instruments
10. failure to follow stated procedures
11. instructions understood incorrectly due to a noise problem.

Inattention and fatigue are two important causes of operator errors. The preventive measures for inattention to be taken into consideration are the installation of attention-getting devices at important points, providing pleasant work places, and eliminating excessive intervals between procedural steps. Similarly, the preventive measures for fatigue are the elimination of awkward positions, excessive times for concentration, stresses due to the environment, tiring mental requirements, and so on.

The failure to note important indications can be avoided by having visual or auditory means to direct the operator's attention to a problem. Furthermore, the use of click-type controls and of those controls that allow adjustments or settings without the necessity for fine movements can avoid the problem of imprecise control settings by the operator.

Remedial measures for controls being switched on in an incorrect sequence are that for critical sequences interlocks should be provided and to ensure that the functional controls are placed in the sequence to be followed for their usage.

The next two causes of error are associated with the use of controls and reading instruments. In the case of mistakes being made in reading instruments, preventive measures such as elimination of visibility problems, the requirement to move the (reader's) body and the avoidance of awkward locations for instruments are important. The problem of mistakes in the use of controls can be avoided by taking actions such as avoidance of excessive force necessitated for use, proximity and similarity of critical controls, and control labels difficult to understand.

Vibration and noise irritation is another cause of operator errors. This can be overcome with preventive measures such as making use of noise-elimination devices and vibration isolators.

With incorporation of the means to ensure that instruments are functioning properly and by providing test and calibration procedures, a cause of operator error (i.e., failure to act at the moment of need due to faulty instruments) can be overcome.

The failure to follow stated procedures is also an important cause for operator errors. Remedial measures for this cause are the avoidance of procedures that are too lengthy, too slow or too fast, etc.

Finally, the last cause (i.e., instructions understood incorrectly due to a noise problem) of operator errors can be overcome by insulating the operator from the noise or reducing the noise at its origin.

SUMMARY

This chapter briefly presents the various aspects of human errors. In the beginning of the chapter various reasons for human errors are listed and their consequences classified into three separate categories. In addition, five ways in which human errors occur are briefly discussed.

Six categories of human errors are explained. These categories are operating errors, assembly errors, design errors, inspection errors, installation errors and maintenance errors.

A formula to calculate human-error probability is presented. Human-error probability estimates for selective tasks are given. The occurrence of human errors in two important areas are discussed. These areas are air traffic control system operations and nuclear power plants. The air traffic control system functions and the tasks of the controller are described.

Finally, four human-error prevention methods are presented. Three of these methods are known as man-machine system analysis, the error-cause removal program and quality control circles. The fourth is not exactly a method but a number of preventive measures to prevent causes of operator errors.

EXERCISES

1. List at least 10 reasons for human errors.
2. Discuss the ways in which human errors occur.
3. Describe the following terms:
   a. operator errors
   b. maintenance errors
   c. design errors.
4. Discuss human errors in production.
5. What are the tasks of an air traffic controller?
6. Discuss the man–machine systems analysis method and its advantages.
7. What are the similarities and differences between the error-cause removal program and quality control circles?
8. Discuss the seven basic elements of the error-cause removal program.

REFERENCES

1. D. Meister, The problem of human-initiated failures, in *Proceedings of the Eighth National Symposium on Reliability and Quality Control*, IEEE, New York, pp. 234–239 (1962).

2. H. L. Williams, Reliability evaluation of the human component in man–machine systems. *Electrical Manufacturing* **4**, 78–82 (1958).
3. B. S. Dhillon, On human reliability — bibliography. *Microelectronics and Reliability* **20**, 371–374 (1980).
4. J. W. Attman, Classification of human errors, in *Proceedings of the Symposium on Reliability of Human Performance in Work* (Edited by W. B. Askren). Rept. AMRL-TR-67-68, Wright-Patterson Air Force Base, Ohio (May 1967).
5. J. I. Cooper, Human-initiated failures and man–function reporting. *IRE Transactions on Human Factors* **10**, 104–109 (1961).
6. J. M. Juran, Operator errors — Time for a new look. *Industrial Quality Control* **1**, 9–11 (1968).
7. R. S. Lincoln, Human factors in the attainment of reliability. *IRE Transactions on Reliability and Quality Control* **11**, 97–103 (1962).
8. R. L. Street, Reducing maintenance by human engineering techniques, in *Proceedings of the Annual Reliability and Maintainability Symposium*, IEEE, New York, pp. 469–471 (1974).
9. K. Inaba and R. Matson, Measurement of human errors with existing data, in *Proceedings of the Seventh Annual Reliability and Maintainability Conference*, IEEE, New York, pp. 301–307 (1968).
10. C. E. Cornell, Minimizing human errors. *Space/Aeronautics* **8** (March 1968).
11. J. A. Kraft, Mitigating of human error through human factors design engineering, in *Proceedings of the Seventh Annual Reliability and Maintainability Conference*, IEEE, New York, p. 300 (1968).
12. L. V. Rigby and A. D. Swain, Effects of assembly error on product acceptability and reliability, in *Proceedings of the Seventh Annual Reliability and Maintainability Conference*, IEEE, New York, pp. 312–319 (1968).
13. L. V. Rigby, Why do people drop things? *Quality Progress* **20**, 16–19 (1973).
14. D. Meister, Reduction of human error, in *Handbook of Industrial Engineering* (Edited by G. Salvendy), pp. 6.2.1–6.2.9. John Wiley & Sons, New York (1982).
15. D. Meister, *Human Factors: Theory and Practice*, pp. 11–56. John Wiley & Sons, New York (1976).
16. A. D. Swain, An error-cause removal program for industry. *Human Factors* **12**, 207–221 (1973).
17. A. Carnino and M. Griffon, Causes of human error, in *High Risk Safety Technology* (Edited by A. E. Green), pp. 171–179. John Wiley & Sons (1982).
18. W. Hammer, *Product Safety Management and Engineering*, pp. 93–107. Prentice-Hall, Englewood Cliffs, New Jersey (1980).
19. A. D. Swain, *Design Techniques for Improving Human Performance in Production*, pp. 5–14. Industrial & Commercial Techniques Ltd., 30-32 Fleet Street, London, EC4 (1973).
20. L. V. Rigby, The nature of human error, in *Proceedings of the Twenty-Fourth Annual Technical Conference of American Society for Quality Control*, IEEE, New York, pp. 457–465 (1970).
21. R. L. McCornack, *Inspector Accuracy: A Study of the Literature*. Report No. SCTM 53-61 (14), Sandia Corporation, Albuquerque, New Mexico (1961).
22. A. E. Green and A. J. Bourne, *Reliability Technology*, p. 22. John Wiley & Sons, London (1972).
23. A. D. Swain and H. E. Guttmann, *Handbook of Human Reliability Analysis with Emphasis on Nuclear Power Plant Application*. Draft Report No. NUREG/CR-1278, United States Nuclear Regulatory Commission, Washington, D.C. (1983).

24. J. W. Danaher, Human error in ATC system operations. *Human Factors* **22**, 535–545 (1980).
25. G. C. Kinney, M. J. Spahn and R. A. Amato, *The Human Element in Air Traffic Control: Observations and Analysis of Performance of Controllers and Supervisors in Providing Air Traffic Control Separation Services.* Report MTR-7655, Produced by METREK Division of the MITRE Corporation (December 1977).
26. D. W. Joos, Z. A. Sabri and A. A. Husseiny, Analysis of gross error rates in operation of commercial nuclear power stations. *Nuclear Engineering and Design* **52**, 265–300 (1979).
27. A. D. Swain and H. E. Guttmann, Human reliability analysis applied to nuclear power, in *Annual Reliability and Maintainability Symposium*, IEEE, New York, pp. 116–119 (1975).
28. R. B. Miller, *A Method for Man-Machine Task Analysis*, Technical Report No. 53-137, Wright Air Development Centre, United States Air Force, Wright–Patterson Air Force Base, Ohio (June 1953).
29. S. Konz, Quality circles: Japanese success story. *Industrial Engineering* **15**, 24–27 (October 1979).

Chapter 5

Human-Reliability Analysis Methods

INTRODUCTION

As numerous systems are interconnected by human links, the predicted reliability of the total system must take into consideration the reliability of the human element. Therefore, it is essential to predict and measure the reliability of humans. In recent years a significant amount of progress has been made in this direction. Many human-performance reliability predictive methods have been developed. The comparative analysis of human reliability models may be found in Ref. [1]. This reference presents the similarities and differences among 22 techniques or procedures for quantitatively predicting the performance of the human element. Basic assumptions such as the following are associated with human reliability prediction methods [2].

1. The task is clearly defined.
2. The approach used can be formulated in proper detail.
3. The approach is divisible into a sequence of subtasks or task components.
4. Individual subtask reliability data are within reach, along with the parameters characterizing the relevant task condition.

This chapter describes selective human-reliability analysis methods.

TECHNIQUE FOR HUMAN ERROR
RATE PREDICTION (THERP)

This technique is a widely known for predicting human error rates [3]. The method is primarily used to evaluate system degradation resulting from human errors in association with factors such as system characteristics influencing people's behavior, operational procedures and the reliability of the equipment. Furthermore, the two basic measures employed by THERP are the probability that an error or group of errors will cause system failure and the probability that an operation will result in an error class j.

THERP is composed of the five steps shown in Fig. 5.1. It is an iterative procedure and its steps are repeated until the system degradation caused by human error is at some tolerable level [4]. It is to be noted here that the steps shown in Fig. 5.1 are not necessarily always repeated in the same order. The steps shown in Figure 5.1 are self-explanatory; only some of them are described here. Steps 2 and 3 are described briefly because of their importance. Description of other steps may be found in Ref. [4]. In step 2, the system and task analysis method is primarily used. Many human-factors specialists are acquainted with this method. All possible human actions and

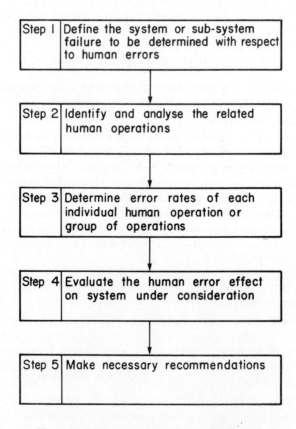

Figure 5.1. Steps associated with THERP.

procedures that can enter into the evaluation process are uncovered with use of a system- and task-analysis method.

The probability tree method usually plays an important role in task analysis. This method is described in the next section. In step 3, the error rates for each human operation associated with system failure are estimated from data available from various sources. Examples of these sources are the AIR Data Store [5] and experimental or other empirical data.

PROBABILITY TREE METHOD

In the THERP technique, the task analysis is usually performed with the aid of the probability tree method. This technique is concerned with representing critical human actions and other events associated with the system under consideration. In addition, diagrammatic task analysis is represented by the branches of the probability tree. Branching limbs of the tree represent outcomes (i.e., success or failure) of each event. Each branch of the tree is assigned an occurrence probability.

Some of the benefits of the probability tree approach [6] are as follows:

1. It is useful in applying predictions of individual error rates and predicts the quantitative effects of errors.
2. It serves as a visibility tool.
3. It can incorporate, with some modifications, factors such as interaction stress, emotional stress and interaction effects.
4. It helps to decrease the probability of errors due to computation because of computational simplification.

The following example describes the basics of this method.

EXAMPLE 5.1

A task of a nuclear power station control room operator is composed of three subtasks, i, j and k. Each of these three subtasks can be either accomplished successfully or unsuccessfully. Furthermore, subtask i is performed before subtask j and subtask j is accomplished before subtask k. In this situation, the unsuccessfully performed subtasks are the only errors that can happen and the performance of one subtask does not affect the performance of the other two subtasks. Develop a probability tree and obtain a probability expression for performing the overall task incorrectly.

The data specified in this example indicate that the operator has to perform subtasks in the order i, j and k irrespective of their outcome (i.e., correct or incorrect). The probability tree for the given data is shown in Fig. 5.2.

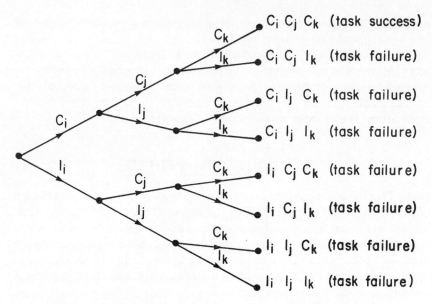

Figure 5.2. Probability tree diagram with subtasks *i*, *j*, and *k*.

The meanings of the symbols used in Fig. 5.2 are defined as: C_i = the subtask *i* is performed correctly, I_i = the subtask *i* is performed incorrectly, C_j = the subtask *j* is performed correctly, I_j = the subtask *j* is performed incorrectly, C_k = the subtask *k* is performed correctly, and I_k = the subtask *k* is performed incorrectly.

From Fig. 5.2, the probability P_{ts} of task success is given by

$$P_{ts} = P(C_i)P(C_j)P(C_k) \ , \tag{5.1}$$

where $P(C_i)$ = the probability of performing subtask *i* correctly; $P(C_j)$ = the probability of performing subtask *j* correctly; and $P(C_k)$ = the probability of performing subtask *k* correctly.

Similarly, from Fig. 5.2, the probability P_{tf} of task failure is given by

$$\begin{aligned} P_{tf} = \ &P(C_i)P(C_j)P(I_k) + P(C_i)P(I_j)P(C_k) \\ &+ P(C_i)P(I_j)P(I_k) + P(I_i)P(C_j)P(C_k) \\ &+ P(I_i)P(C_j)P(I_k) + P(I_i)P(I_j)P(C_k) + P(I_i)P(I_j)P(I_k) \ , \end{aligned} \tag{5.2}$$

where $P(I_i)$ = the probability of performing subtask i incorrectly, $P(I_j)$ = the probability of performing subtask j incorrectly, and $P(I_k)$ = the probability of performing subtask k incorrectly.

Since $P(I_k) = 1 - P(C_k)$, $P(I_j) = 1 - P(C_j)$ and $P(I_i) = 1 - P(C_i)$, Eq. (5.2) reduces to

$$P_{\text{tf}} = 1 - P(C_i)P(C_j)P(C_k) \ . \tag{5.3}$$

As expected, Eqs. (5.1) and (5.3) indicate that the task will only be performed successfully if all of the subtasks i, j and k are accomplished correctly.

EXAMPLE 5.2

If the last subtask k in Example 5.1 is eliminated, develop a probability tree and an expression for correctly performing the task. Calculate the probability of performing the task incorrectly if the probabilities of performing subtasks i and j correctly are 0.8 and 0.7, respectively.

As there are only two subtasks, the probability tree of Fig. 5.2 reduces to the one shown in Fig. 5.3. With the aid of this diagram, the probability P_{ts} of task success is

$$P_{\text{ts}} = P(C_i)P(C_j) \ . \tag{5.4}$$

Similarly, with the aid of the diagram shown in Fig. 5.3, the probability P_{tf} of task failure is

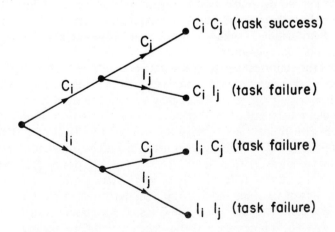

Figure 5.3. Probability tree diagram with subtasks i and j.

$$P_{tf} = P(C_i)P(I_j) + P(I_i)P(C_j) + P(I_i)P(I_j) \ . \tag{5.5}$$

Since $P(I_i) = 1 - P(C_i)$ and $P(I_j) = 1 - P(C_j)$, Eq. (5.5) reduces to

$$P_{tf} = 1 - P(C_i)P(C_j) \ . \tag{5.6}$$

By substituting $P(C_i) = 0.8$ and $P(C_j) = 0.7$ in Eq. (5.6), the probability of performing the task incorrectly is

$$P_{tf} = 1 - (0.8)(0.7)$$
$$= 0.44 \ .$$

PONTECORVO'S METHOD OF PREDICTING HUMAN RELIABILITY

This method is useful for obtaining reliability estimates of task performance [7]. The method is concerned with obtaining estimates of reliability of separate and discrete subtasks having no correct reliability figures. The total task reliability estimate is obtained by combining these estimates. To quantitatively assess the interaction of men and machines, the method is applied during the initial phases of design. In addition, this approach is applicable to determining the performance of a single person acting alone. Pontecorvo's approach is composed of six steps. All of these steps are shown in Fig. 5.4. The steps shown in this figure are described below. Step 1 is concerned with the identification of tasks to be performed. These tasks should be identified at a gross level; in other words, one complete operation is to be represented by each task. Once the tasks are established the next logical step (step 2) is concerned with the identification of subtasks of each task (those subtasks which are necessary for task completion).

In step 3 the empirical performance data are collected. These data should be subject to those environments under which the subtask is to be accomplished. Empirical performance data may be obtained from various sources, for example, the experimental literature and in-house operations. Step 4 is concerned with rating each subtask according to its level of difficulty or potential for error. A 10-point scale is used for judging the subtask rate. The scale varies from least error to most error. The next step (step 5) is concerned with predicting subtask reliability. In order to predict subtask reliability the judged ratings of the data and the empirical data are expressed in the form of a straight line. The regression line is tested for goodness of fit.

The last step of this approach is concerned with determining the task reliability. The task reliability is given by the product of subtask reliabilities. The

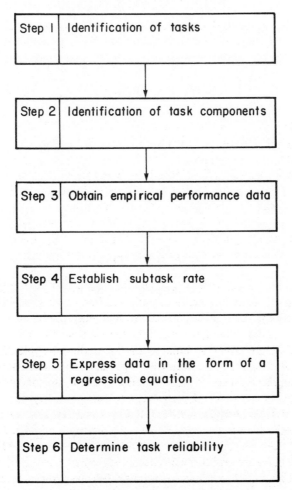

Figure 5.4. Steps of Pontecorvo's approach.

above approach is used to determine the performance of a single individual acting alone. However, in a situation where a backup person is available, the probability of the task being performed correctly (i.e., the task reliability) would be greater. Nevertheless, the redundant backup individual may not be available all of the time. Therefore, according to Ref. [7], when two persons are working together to accomplish a task, the overall reliability is given by

$$R_0 = [\{1 - (1 - r)^2\}t_a + rt_u]/(t_a + t_u) \ , \tag{5.7}$$

where t_a denotes the percentage of time the backup human is available, t_u denotes the percentage of time the backup human is unavailable, r denotes the reliability of the single person, and

$$t_a + t_u = 100\% \ . \tag{5.8}$$

EXAMPLE 5.3

Two independent persons are working together to perform a maintenance task. The reliability of each person is 0.85. Only 70% of time is the backup person available. In other words, 30% of time the backup person is not available. Calculate the reliability of performing the maintenance task correctly.

In this example the data are specified for the following elements of Eq. (5.7):

$$r = 0.85 \ , \qquad t_a = 0.7 \qquad \text{and} \qquad t_u = 0.3 \ .$$

Substituting the above data into Eq. (5.7) results in

$$R_0 = [\{1 - (1 - 0.85)^2\}0.7 + (0.85)(0.3)]/(0.7 + 0.3)$$
$$= 0.93925 \ .$$

The probability of performing the maintenance task correctly is 0.93925.

THE THROUGHPUT RATIO METHOD

This is a reliability-oriented predictive technique developed by the Navy Electronics Laboratory Center [1]. The ratio determines the operability of man–machine interfaces or stations.

The typical example is the control panel. According to Ref. [1], the definition of operability is the extent to which the man–machine station performance fulfills the design expectation for the station in question.

The term "throughput" implies transmission, because the ratio is in terms of responses or items per unit time emitted by the operator. The throughput ratio in percentage is defined by

$$P_{mo} = \left(\frac{\alpha}{\beta} - E_c \right) \times 100 \ , \tag{5.9}$$

where α represents the throughput items generated per unit time; β represents the throughput items to be generated per unit time in order to satisfy design expectation; P_{mo} represents the man–machine operability; and E_c represents a correction factor, namely correction for error or out-of-tolerance output.

The correction factor, E_c, is defined by

$$E_c = \prod_{i=1}^{2} X_i \, , \tag{5.10}$$

$$X_1 \equiv \frac{T_1}{T_2} \frac{\alpha}{\beta} \, , \tag{5.11}$$

$$X_2 \equiv \frac{\alpha}{\beta} \frac{T_1}{T_2} P_1^2 P_2 \, , \tag{5.12}$$

where P_1 is the probability that the error will not be detected by the operator; P_2 is the probability of function failure due to an error; T_1 are the trials in which the control-display operation is carried out incorrectly; and T_2 is the total of trials in which the control-display operation is carried out.

This throughput ratio may be used for purposes such as the following.

1. To demonstrate system acceptability. In this respect it serves as a measurement tool.
2. Redesign of design that was already evaluated. In this respect it is used to correct human engineering discrepancies.
3. To establish the feasibility of system.
4. To make comparisons of alternative design operabilities.

EXAMPLE 5.4

Calculate the value of the correction factor E_c, if the values of T_1, T_2, α, β, P_1 and P_2 are as follows:

$$T_1 = 3 \, , \quad T_2 = 12 \, , \quad \alpha = 6 \, , \quad \beta = 12 \, , \quad P_1 = 0.3 \, , \quad P_2 = 0.6 \, .$$

Substituting the above data into Eqs. (5.10)–(5.12) we get

$$E_c = (0.125)(0.0068) = 0.0008 \, ,$$

$$X_1 = \frac{3}{12} \frac{6}{12} = 0.125 \, ,$$

$$X_2 = \frac{6}{12} \frac{3}{12} (0.3)^2 (0.6) = 0.0068 \, .$$

The value of the correction factor is 0.0008.

EXAMPLE 5.5

With the aid of data specified in Example 5.4 calculate the value of the throughput ratio.

By substituting the given data into Eq. (5.9) we obtain

$$P_{mo} = \left(\frac{6}{12} - 0.0008 \right) \times 100$$

$$= 49.92\% \ .$$

Thus the value of the man–machine operability is 49.92%.

PERSONNEL RELIABILITY INDEX

This index was developed to provide feedback on the technical proficiency of the electronic maintenance manpower of the United States Navy [8]. The index is based on nine job factors; it makes use of information such as instruction, personnel relationships, equipment operation, equipment inspection, electrocognition, electronic circuit analysis, using reference materials, electrosafety and electrorepair. Various types of activities are associated with these factors. For example, the activities associated with electrorepair, equipment inspection, instruction, electrosafety, personnel relationships and equipment operation are the repairing of equipment in the shop; supervising and performing inspections of electronic equipment; teaching other people with respect to maintenance, operation and inspection of electronic equipment; making use of safety precautions on oneself and on equipment; supervision of electronic equipment maintenance, operation and inspection; and operating electrical and electronics test equipment, etc.

Activities such as making out reports and using supporting reference materials are associated with the using-reference-materials factor. Examples of activities associated with the electrocognition factor are the use of electronic maintenance reference materials and maintenance and troubleshooting of electronic equipment. Finally, activities such as preparing failure reports, keeping maintenance usage data and understanding electronic circuitry principles are associated with the electronic circuit analysis factor.

For each of the above job factors the data are collected from maintenance supervisors over the period of 2 months. These data are concerned with the number of uncommonly effective and uncommonly ineffective performances by maintenance personnel. With the aid of such data for each job factor, the value of the following ratio, R, is computed:

$$R = \frac{\Sigma \alpha}{\Sigma \alpha + \Sigma \beta} \ , \tag{5.13}$$

where α are the uncommonly effective behaviours, and β are the uncommonly ineffective behaviours. The value of R varies between 0 and 1. The total effectiveness value, E, for a maintenance person is given by

$$E = R_1 R_2 R_3 R_4 R_5 R_6 R_7 R_8 R_9 \ , \tag{5.14}$$

where R_i is the ith factor ratio value (i.e., reliability) for $i = 1, 2, \ldots, 9$.

The anticipated uses of the personnel reliability index are in design analysis, manpower training and selection, etc. Finally, it can be added that the method is based on avionics maintenance activities, but it should be possible to use it for any kind of maintenance.

BLOCK DIAGRAM METHOD

This method is used to compute the reliability of a parallel system. In this system it is assumed that all the units are active and at least one unit must function normally for system success. The system fails when all of its units fail. Each unit of the system may fail due to the occurrence of a hardware failure or a noncritical human error. Furthermore, human errors are classified into two categories, critical and noncritical human errors. A critical human error causes all of the system units to fail simultaneously, whereas only one unit fails due to a noncritical human error. The method assumes that the system units fail independently. The method is demonstrated with the aid of Fig. 5.5. The diagram shown in this figure is composed of n units. Each unit's failure probabilities are separated into probabilities of hardware failures and human errors. In addition, all of the system units fail due to the occurrence of a critical human error. A hypothetical unit representing critical human errors is also shown in the diagram. A typical example of a critical human error is fire due to people in a room containing the parallel system. The reliability, R_p, of the parallel system shown in Fig. 5.5 is given by

$$R_p = \left[1 - \prod_{i=1}^{n} \{1 - (1 - F_i)(1 - f_i)\} \right] (1 - f_c) \ , \tag{5.15}$$

where n is the number of active units, F_i is the hardware failure probability of ith unit for $i = 1, 2, 3, \ldots, n$, f_i is the failure probability of ith unit due to noncritical human errors for $i = 1, 2, 3, \ldots, n$, and f_c is the failure probability of the parallel system due to critical human errors.

Since $R_i = 1 - F_i$ and $r_i = 1 - f_i$, Eq. (5.15) simplifies to

$$R_p = \left[1 - \prod_{i=1}^{n} (1 - R_i r_i) \right] (1 - f_c) \ , \tag{5.16}$$

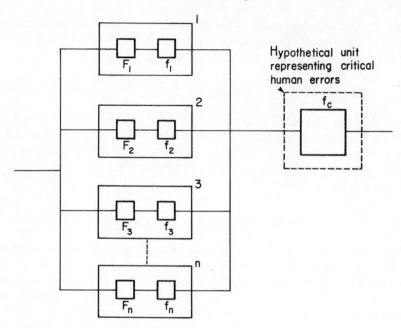

Figure 5.5. A parallel system with critical and noncritical human errors.

where R_i is the hardware reliability of ith unit and r_i is the reliability of ith unit with respect to noncritical human errors. For identical units, Eq. (5.16) reduces to

$$R_p = [1 - (1 - Rr)^n](1 - f_c) \ . \tag{5.17}$$

For the specified values of R, r, f_c and n the plots of Eq. (5.17) are shown in Fig. 5.6. As expected, the system reliability decreases with increasing values of f_c. In addition, the relative increase in system reliability is higher for adding the third unit than the fourth one.

EXAMPLE 5.6

A parallel system is composed of two independent and identical active units. Each unit may fail due to a hardware failure or a noncritical human error. The system may fail due to the occurrence of a critical human error.

The probability of occurrence of a critical human error is 0.01. In addition, the probabilities of occurrence of a hardware failure and a noncritical human error are 0.05 and 0.02, respectively. Calculate the system reliability.

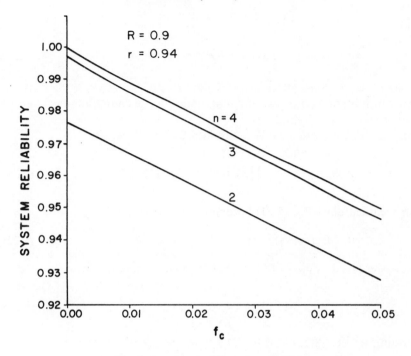

Figure 5.6. Plots of equation (5.17).

With the aid of specified data the values of R, r, n and f_c are as follows: $R = 1 - 0.05 = 0.95$, $r = 1 - 0.02 = 0.98$, $f_c = 0.01$, and $n = 2$. By substituting the above values in Eq. (5.17) we get

$$R_p = [1 - (1 - (0.95)(0.98))^2](1 - 0.01)$$

$$= 0.9853 \ .$$

Thus the system reliability is 0.9853.

Time-Dependent Analysis

When hardware failure and critical and noncritical human error rates are constant, the time (t) dependent equations for R, r and f_c are

$$R(t) = e^{-\lambda_R t} \ , \tag{5.18}$$

$$r(t) = e^{-\lambda_r t} \ , \tag{5.19}$$

and

$$f_c(t) = 1 - e^{-\lambda_c t} , \tag{5.20}$$

where λ_R is the constant hardware failure rate of a unit, λ_r is the constant noncritical human error rate of a unit, and λ_c is the constant critical human error rate of the system.

By substituting Eqs. (5.18)–(5.20) into Eq. (5.16) we get

$$R_p(t) = \left[1 - \prod_{i=1}^{n} (1 - e^{(-\lambda_{Ri} + \lambda_{ri})t}) \right] e^{-\lambda_c t} . \tag{5.21}$$

For identical units Eq. (5.21) becomes

$$R_p(t) = [1 - (1 - e^{-(\lambda_R + \lambda_r)t})^n] e^{-\lambda_c t} . \tag{5.22}$$

The system mean time to failure (MTTF) can be obtained from

$$\text{MTTF} = \int_0^\infty R_p(t) \, dt . \tag{5.23}$$

Substituting Eq. (5.22) into Eq. (5.23) and integrating leads to

$$\text{MTTF} = \frac{1}{\lambda_c} - \sum_{i=0}^{n} \binom{n}{i} (-1)^{n-i} \frac{1}{(n-i)(\lambda_R + \lambda_r) + \lambda_c} . \tag{5.24}$$

EXAMPLE 5.7

The following data are specified for the parallel system of Example 5.6: $\lambda_R = 0.004$ failure/hr, $\lambda_r = 0.0002$ error/hr, and $\lambda_c = 0.0001$ error/hr. Calculate the system reliability for a 200 hr mission and calculate the mean time to failure.

With the aid of Eq. (5.22) and the given data we get

$$R_p(200) = [1 - (1 - e^{-(0.004 + 0.0002)(200)})^2] e^{-(0.0001)(200)}$$

$$= 0.6636 .$$

Similarly, for $n = 2$, the mean time to failure from Eq. (5.24) is

$$\text{MTTF} = \frac{1}{\lambda_c} - \sum_{i=0}^{2} \binom{2}{i} (-1)^{2-i} \frac{1}{(2-i)(\lambda_R + \lambda_r) + \lambda_c} \tag{5.25}$$

$$= \frac{1}{\lambda_c} - \frac{1}{2(\lambda_R + \lambda_r) + \lambda_c} + \frac{2}{\lambda_R + \lambda_r + \lambda_c} - \frac{1}{\lambda_c}$$

$$= \frac{2}{\lambda_R + \lambda_r + \lambda_c} - \frac{1}{2\lambda_R + 2\lambda_r + \lambda_c} . \tag{5.26}$$

Utilizing the given data and inserting into Eq. (5.26) we get

$$\text{MTTF} = \frac{2}{(0.004) + (0.0002) + 0.0001} - \frac{1}{2(0.004) + 2(0.0002) + 0.0001}$$

$$= \frac{2}{0.0043} - \frac{1}{0.0085} = 347.47 \text{ hr} .$$

Thus the parallel-system expected time to failure is 347.47 hr.

SUMMARY

This chapter presents selected human-reliability analysis methods. Methods covered in this chapter are

1. technique for human error rate prediction (THERP)
2. probability tree method
3. Pontecorvo's method
4. throughput ratio method
5. personnel reliability index
6. block diagram method.

The technique for human-error rate prediction is briefly discussed and associated steps are outlined. The probability tree method is described with the aid of two solved examples. The benefits of the method are listed. Pontecorvo's method is concerned with obtaining the reliability estimate of task performance. The method is composed of six steps. All of these steps are described briefly.

A formula to predict the reliability of two persons working together to accomplish a task is given. The throughput ratio method is presented with the aid of a numerical example. Application areas of the method are listed.

The personnel reliability index method was developed to provide feedback on the technical proficiency of electronic maintenance manpower of the United States Navy. This method makes use of information on nine job factors: instruction, personnel relationships, equipment operation, equipment inspection, electrocognition, electronic circuit analysis, using reference materials, electrosafety and electrorepair. All these factors are briefly discussed and equations associated with the method are given.

Finally, the block diagram approach is concerned with evaluating the reliability of a parallel system with human errors. Human errors are separated into two classes, i.e., critical and noncritical human errors. An expression to predict the mean time to failure of the parallel system is developed. Two numerical examples are presented.

EXERCISES

1. Compare the technique for human-error rate prediction with Pontecorvo's method. List the drawbacks of both techniques.
2. What are the advantages and application areas of THERP?
3. Two independent maintenance workers are working together to accomplish a task. The probability of success of performing the task successfully by each person is 0.95. However, the backup worker (i.e., second worker) is available only 60% of the time. Compute the probability of performing the task successfully.
4. What are the differences between the block diagram method and the other methods discussed in this chapter?
5. Discuss the historical development of the following two methods:
   a. the throughput ratio method;
   b. the personnel reliability index.
6. Prove that the mean time to failure of a three-unit parallel system is given by

$$\text{MTTF} = \frac{1}{3\lambda_R + 3\lambda_r + \lambda_c} - \frac{3}{2\lambda_R + 2\lambda_r + \lambda_c} + \frac{3}{\lambda_R + \lambda_r + \lambda_c} , \quad (5.27)$$

where λ_c is the constant critical human error rate of the system; λ_r is the constant noncritical human-error rate of a unit; and λ_R is the constant hardware failure rate of a unit.

7. Discuss the assumptions associated with the parallel system of Exercise 6.

REFERENCES

1. D. Meister, *Comparative Analysis of Human Reliability Models*, Report No. AD 734 432, 1971, p. 481. Available from the National Technical Information Service, Springfield, Virginia 22151.
2. J. Rasmussen, The role of the man–machine interface in systems reliability, in *Proceedings of the NATO Generic Conference*, Liverpool, UK, Noordhoff International Pub. Co., Leyden, The Netherlands, pp. 315–323 (1973).
3. A. D. Swain, *A Method for Performing a Human-Factors Reliability Analysis*. Report SCR-685, Sandia Corporation, Albuquerque, New Mexico (August 1963).
4. D. Meister, Methods of predicting human reliability in man–machine systems. *Human Factors* **6**, 621–646 (1964).
5. S. J. Munger, R. W. Smith and D. P. Payne, *An Index of Electronic Equipment Operability: Data Store*, AIR-C43-1/62-RP(1), American Institute for Research, Pittsburgh, Pennsylvania (January 1962).
6. B. S. Dhillon and C. Singh, *Engineering Reliability: New Techniques and Applications*. John Wiley & Sons, New York (1981).
7. A. B. Pontecorvo, A method of predicting human reliability, in *Proceedings of Fourth Annual Reliability and Maintainability Conference*, pp. 337–342. Spartan Books, Washington, D.C. (1965).
8. A. I. Siegel and P. J. Federman, *Development of Performance Evaluative Measures*. Report No. 7071-2, Contract N0014-67-00107, Office of Naval Research, United States Navy, Washington, D.C. (September 1970).

Chapter 6

Reliability Evaluation of Systems with Human Errors

INTRODUCTION

Humans interact with engineering systems in many ways. Examples of interactions may be seen each day in nuclear power plants, computer operation rooms, cockpits of airplanes, and so on. Human error is a very controversial and important topic in reliability engineering. Not all human errors necessarily cause system failures. In addition, some human errors cause more "significant" failures than others.

According to the findings of some researchers, abut 20–30% of system failures are directly or indirectly related to human error. Human error in engineering systems is mainly due to factors such as maintenance errors, misinterpretation of instruments and incorrect actions.

This chapter presents mathematical models used to perform reliability and availability analysis of various types of engineering systems with human errors. Markov and supplementary variable techniques [1] were used to develop these models. Various plots are presented in this chapter. These plots were developed for arbitrarily specified values of parameters involved because the intention here is to show plots without taking into consideration their real-life significance. Consequently, the specified values of the human error rate in some plots is greater than the unit hardware failure rate.

RELIABILITY EVALUATION OF REDUNDANT SYSTEMS WITH HUMAN ERROR

Redundancy plays an important role in increasing systems' reliability. There are several types of redundant configurations used to enhance system reliability. The parallel, standby and k-out-of-n units configurations are some of the widely used redundant configurations. This section is therefore concerned with parallel, standby, and k-out-of-n units redundant configurations. In the usual reliability analysis of these redundant systems only hardware failures are considered, and the occurrence of human error is neglected. With the increased interaction between humans and machines, human error has become one of the important failure factors. According to one study [2], about 10–15% of the total system failures are due to human errors. Some of the causes of human errors are lack of good job environments, inadequate tools and operating procedures for the operating personnel, poor training or skill of the operating personnel, etc. Thus the probability of redundant system failure due to human error is a vital parameter. Realistic system reliability analysis must include human and hardware failure aspects; this section is therefore confined to the analysis of redundant systems subject to hardware failures and human errors [3,4].

Analysis of a Parallel System

This section presents four Markov models pertaining to nonrepairable active parallel systems subject to hardware failure and human error [3]. Models I, II, III and IV are concerned with two-unit, three-unit, four-unit and n-unit parallel systems, respectively. Each unit of parallel systems may fail either due to a hardware failure or a human error.

The following assumptions are associated with all four models.

1. Each parallel system is composed of identical units.
2. Failures are statistically independent.
3. Each unit's hardware failure and human error rates are constant.
4. Failed units are never repaired.
5. Each unit's human and hardware failures can be separated.
6. All units of each system operate simultaneously.
7. Numerals in the boxes of Fig. 6.1 denote corresponding states.

In addition, symbols such as the following were used to develop all four models: n is the number of units in the parallel system; $P_i(t)$ is the probability that the system is in state i at time t for $i = 0,1,2,\ldots,2n$ (i.e., for $n = 2,3,4,\ldots$); λ_1 is the constant hardware failure rate of a unit; λ_h is the constant human error rate of a unit; and s is the Laplace transform variable.

Model I. This model represents a two-unit active parallel system. The resulting diagram for $n = 2$ in Fig. 6.1 represents the system transition diagram for model I.

For $n = 2$, the following system of differential equations is associated with Fig. 6.1 (a detailed simple example of setting up differential equations with a Markov system transition diagram is given in chapter 3):

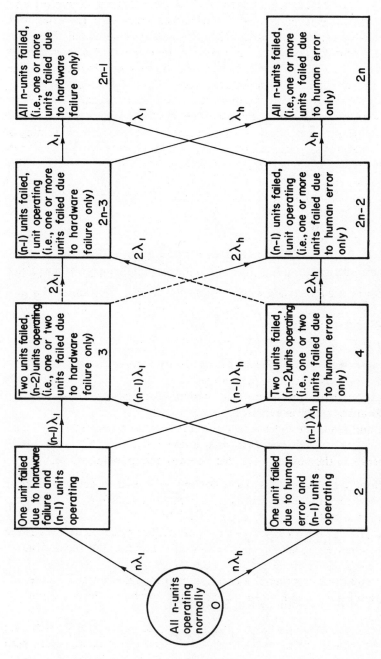

Figure 6.1. System transition diagram for an *n*-unit parallel system.

$$\frac{dP_0(t)}{dt} + 2(\lambda_1 + \lambda_h)P_0(t) = 0 \; , \tag{6.1}$$

$$\frac{dP_1(t)}{dt} + (\lambda_1 + \lambda_h)P_1(t) = P_0(t)2\lambda_1 \; , \tag{6.2}$$

$$\frac{dP_2(t)}{dt} + (\lambda_1 + \lambda_h)P_2(t) = P_0(t)2\lambda_h \; , \tag{6.3}$$

$$\frac{dP_3(t)}{dt} = P_1(t)\lambda_1 + P_2(t)\lambda_1 \; , \tag{6.4}$$

$$\frac{dP_4(t)}{dt} = P_1(t)\lambda_h + P_2(t)\lambda_h \; . \tag{6.5}$$

At $t = 0$, $P_0(0) = 1$, $P_1(0) = P_2(0) = P_3(0) = P_4(0) = 0$. With the aid of Laplace transforms, from Eqs. (6.1)–(6.5) the resulting state probability equations are

$$P_0(t) = e^{-2At} \tag{6.6}$$

where $A = \lambda_1 + \lambda_h$,

$$P_1(t) = \lambda_1 B \tag{6.7}$$

where $B = (2/A)(e^{-At} - e^{-2At}) \; ,$

$$P_2(t) = \lambda_h B \tag{6.8}$$

and

$$P_3(t) = \lambda_1 C \tag{6.9}$$

where $C = (1/A)(1 - e^{-At})^2$, and

$$P_4(t) = \lambda_h C \; . \tag{6.10}$$

Reliability of the two-unit system is given by

$$R(t) = P_0(t) + P_1(t) + P_2(t)$$
$$= 1 - (1 - e^{-At})^2 \; . \tag{6.11}$$

The mean time to failure of the system [1] is given by

$$\text{MTTF} = \int_0^\infty R(t)\, dt = \frac{1}{A} + \frac{1}{2A} \ . \tag{6.12}$$

EXAMPLE 6.1

A parallel system is composed of two identical and independent active units. Each unit may fail due to a hardware failure or a human error. Unit constant hardware failure and human-error rates are 0.004 failure/hr and 0.0005 error/hr, respectively. Calculate the two-unit system mean time to failure.

By substituting the specified data into Eq. (6.12), we get

$$\begin{aligned}
\text{MTTF} &= \frac{1}{A} + \frac{1}{2A} = \frac{1}{\lambda_1 + \lambda_h} + \frac{1}{2(\lambda_1 + \lambda_h)} \\
&= \frac{1}{(0.004) + (0.0005)} + \frac{1}{2(0.004 + 0.0005)} \\
&= 333.33 \text{ hr} \ .
\end{aligned}$$

Thus the two-unit parallel-system mean time to failure is 333.33 hr.

The probability that at time t at least one unit is in failed condition due to human error is given by

$$P_h(t) = P_2(t) + P_4(t) \ . \tag{6.13}$$

Plots of the above equation are shown in Fig. 6.2 for specific values of λ_h and λ_1. Figure 6.2 shows that $P_h(t)$ increases for the corresponding increasing values of t and λ_h.

The probability that at time t at least one unit is in failed condition due to hardware failure is given by

$$P(t) = P_1(t) + P_3(t) \ . \tag{6.14}$$

Plots of the above equation are shown in Fig. 6.3 for specific values of λ_h and λ_1. This figure shows that $P(t)$ increases for the corresponding increasing values of t but decreases for the increasing values of λ_h.

Model II. Model II is concerned with a three-unit active parallel system. Figure 6.1 yields the system transition diagram for model II when $n = 3$.

The system of differential equations associated with model II is

Human Reliability

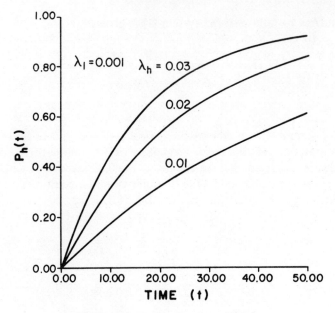

Figure 6.2. $P_h(t)$ plots for model I.

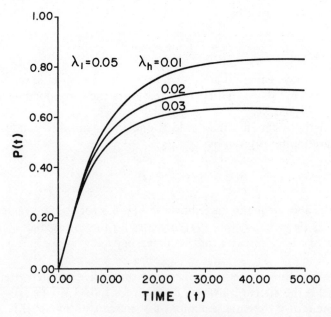

Figure 6.3. $P(t)$ plots for model I.

$$\frac{\mathrm{d}P_0(t)}{\mathrm{d}t} + 3(\lambda_1 + \lambda_h)P_0(t) = 0 \ , \tag{6.15}$$

$$\frac{\mathrm{d}P_1(t)}{\mathrm{d}t} + 2(\lambda_1 + \lambda_h)P_1(t) = P_0(t)3\lambda_1 \ , \tag{6.16}$$

$$\frac{\mathrm{d}P_2(t)}{\mathrm{d}t} + 2(\lambda_1 + \lambda_h)P_2(t) = P_0(t)3\lambda_h \ , \tag{6.17}$$

$$\frac{\mathrm{d}P_3(t)}{\mathrm{d}t} + (\lambda_1 + \lambda_h)P_3(t) = P_1(t)2\lambda_1 + P_2(t)2\lambda_1 \ , \tag{6.18}$$

$$\frac{\mathrm{d}P_4(t)}{\mathrm{d}t} + (\lambda_1 + \lambda_h)P_4(t) = P_1(t)2\lambda_h + P_2(t)2\lambda_h \ , \tag{6.19}$$

$$\frac{\mathrm{d}P_5(t)}{\mathrm{d}t} = P_3(t)\lambda_1 + P_4(t)\lambda_1 \ , \tag{6.20}$$

$$\frac{\mathrm{d}P_6(t)}{\mathrm{d}t} = P_3(t)\lambda_h + P_4(t)\lambda_h \ . \tag{6.21}$$

At $t = 0$, $P_0(0) = 1$, and all other initial condition probabilities are equal to zero.

With the aid of Laplace transforms, we find from Eqs. (6.15)–(6.21) the resulting state probability equations:

$$P_0(t) = e^{-3At} \tag{6.22}$$

where $A = \lambda_1 + \lambda_h$,

$$P_1(t) = \lambda_1 D \tag{6.23}$$

where $D = (3/A)(e^{-2At} - e^{-3At})$,

$$P_2(t) = \lambda_h D \tag{6.24}$$

and

$$P_3(t) = \lambda_1 E \tag{6.25}$$

where $E = (3/A)(e^{-At} - 2e^{-2At} + e^{-3At})$,

$$P_4(t) = \lambda_h E \tag{6.26}$$

and

$$P_5(t) = \lambda_1 F \qquad (6.27)$$

where $F = (1/A)(1 - e^{-At})^3$, and finally

$$P_6(t) = \lambda_h F , \qquad (6.28)$$

Reliability of the system is given by

$$R(t) = \sum_{i=0}^{4} P_i(t) = 1 - (1 - e^{-At})^3 . \qquad (6.29)$$

The mean time to system failure is given by

$$\text{MTTF} = \int_0^\infty R(t)\, dt = \frac{1}{A} + \frac{1}{2A} + \frac{1}{3A} . \qquad (6.30)$$

EXAMPLE 6.2

Assume that the parallel system of Example 6.1 contains three units instead of only two units. Compute the three-unit system mean time to failure.

In this example the unit hardware failure rate $\lambda_1 = 0.004$ failure/hr and the unit human-error rate $\lambda_h = 0.0005$ error/hr. By substituting these data into Eq. (6.30) we get

$$\text{MTTF} = \frac{1}{A} + \frac{1}{2A} + \frac{1}{3A} = \frac{1}{(\lambda_1 + \lambda_h)} + \frac{1}{2(\lambda_1 + \lambda_h)} + \frac{1}{3(\lambda_1 + \lambda_h)}$$

$$= \frac{1}{(0.004 + 0.0005)} + \frac{1}{2(0.004 + 0.0005)} + \frac{1}{3(0.004 + 0.0005)}$$

$$= 407.40 \text{ hr} .$$

Thus the three-unit parallel-system mean time to failure is 407.40 hr.

The probability that at time t at least one unit is in failed condition due to human error is given by

$$P_h(t) = P_2(t) + P_4(t) + P_6(t) . \qquad (6.31)$$

The plots of this equation are shown in Fig. 6.4 for given values of λ_h and λ_1. These plots show that $P_h(t)$ increases for the corresponding increasing values of t and λ_h.

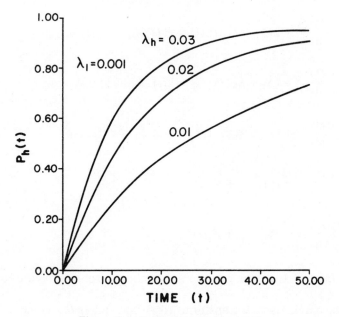

Figure 6.4. $P_h(t)$ plots for model II.

The probability that at time t at least one unit is in failed condition due to hardware failure is given by

$$P(t) = P_1(t) + P_3(t) + P_5(t) \ . \tag{6.32}$$

Model III. This model represents a four-unit active parallel system. In Fig. 6.1, the resulting figure for $n = 4$ represents the system transition diagram for model III. The system of differential equations associated with model III is

$$\frac{dP_0(t)}{dt} + 4(\lambda_1 + \lambda_h)P_0(t) = 0 \ , \tag{6.33}$$

$$\frac{dP_1(t)}{dt} + 3(\lambda_1 + \lambda_h)P_1(t) = P_0(t)4\lambda_1 \ , \tag{6.34}$$

$$\frac{dP_2(t)}{dt} + 3(\lambda_1 + \lambda_h)P_2(t) = P_0(t)4\lambda_h \ , \tag{6.35}$$

$$\frac{dP_3(t)}{dt} + 2(\lambda_1 + \lambda_h)P_3(t) = P_1(t)3\lambda_1 + P_2(t)3\lambda_1 \ , \tag{6.36}$$

$$\frac{\mathrm{d}P_4(t)}{\mathrm{d}t} + 2(\lambda_1 + \lambda_h)P_4(t) = P_1(t)3\lambda_h + P_2(t)3\lambda_h \ , \tag{6.37}$$

$$\frac{\mathrm{d}P_5(t)}{\mathrm{d}t} + (\lambda_1 + \lambda_h)P_5(t) = P_3(t)2\lambda_1 + P_4(t)2\lambda_1 \ , \tag{6.38}$$

$$\frac{\mathrm{d}P_6(t)}{\mathrm{d}t} + (\lambda_1 + \lambda_h)P_6(t) = P_3(t)2\lambda_h + P_4(t)2\lambda_h \ , \tag{6.39}$$

$$\frac{\mathrm{d}P_7(t)}{\mathrm{d}t} = P_5(t)\lambda_1 + P_6(t)\lambda_1 \ , \tag{6.40}$$

$$\frac{\mathrm{d}P_8(t)}{\mathrm{d}t} = P_5(t)\lambda_h + P_6(t)\lambda_h \ . \tag{6.41}$$

At $t = 0$, $P_0(0) = 1$, and all other initial condition probabilities are equal to zero.

With the aid of Laplace transforms from Eqs. (6.33)–(6.41) the resulting state probability equations are

$$P_0(t) = e^{-4At} \tag{6.42}$$

where $A = \lambda_1 + \lambda_h$,

$$P_1(t) = \lambda_1 G \tag{6.43}$$

where $G = (4/A)(e^{-3At} - e^{-4At})$,

$$P_2(t) = \lambda_h G \tag{6.44}$$

and

$$P_3(t) = \lambda_1 H \tag{6.45}$$

where $H = (6/A)(e^{-2At} - 2e^{-3At} + e^{-4At})$,

$$P_4(t) = \lambda_h H \tag{6.46}$$

and

$$P_5(t) = \lambda_1 I \tag{6.47}$$

where $I = (4/A)(e^{-At} - 3e^{-2At} + 3e^{-3At} - e^{-4At})$,

$$P_6(t) = \lambda_h I \qquad (6.48)$$

and

$$P_7(t) = \lambda_l J \qquad (6.49)$$

where $J = (1/A)(1 - e^{-At})^4$, and finally

$$P_8(t) = \lambda_h J . \qquad (6.50)$$

Reliability of the system is given by

$$R(t) = \sum_{i=0}^{6} P_i(t) = 1 - (1 - e^{-At})^4 . \qquad (6.51)$$

The mean time to failure of the system is given by

$$\text{MTTF} = \int_0^\infty R(t)\, \mathrm{d}t = \frac{1}{A} + \frac{1}{2A} + \frac{1}{3A} + \frac{1}{4A} . \qquad (6.52)$$

EXAMPLE 6.3

A parallel system contains four identical and independent active units. Each unit may fail due to a hardware failure or a human error. In addition, the unit constant hardware failure and human-error rates are 0.004 failure/hr and 0.0005 error/hr, respectively. Determine the parallel-system mean time to failure.

In this example the following values are specified for λ_l and λ_h: $\lambda_l = 0.004$ failure/hr, and $\lambda_h = 0.0005$ error/hr.

By substituting the specified data into Eq. (6.52), the resulting value of the mean time to failure is

$$
\begin{aligned}
\text{MTTF} &= \frac{1}{(\lambda_l + \lambda_h)} + \frac{1}{2(\lambda_l + \lambda_h)} + \frac{1}{3(\lambda_l + \lambda_h)} + \frac{1}{4(\lambda_l + \lambda_h)} \\
&= \frac{1}{(0.004 + 0.0005)} + \frac{1}{2(0.004 + 0.0005)} + \frac{1}{3(0.004 + 0.0005)} \\
&\quad + \frac{1}{4(0.004 + 0.0005)} \\
&= 462.96 \text{ hr} .
\end{aligned}
$$

The probability that at time t at least one unit is in failed condition due to human error is given by

$$P_h(t) = P_2(t) + P_4(t) + P_6(t) + P_8(t) \ . \tag{6.53}$$

Similarly, the probability that at time t at least one unit is in failed condition due to hardware failure is given by

$$P(t) = P_1(t) + P_3(t) + P_5(t) + P_7(t) \ . \tag{6.54}$$

Model IV. This model is concerned with the analysis of an n-unit active parallel system subject to hardware failure and human error. The system transition diagram is shown in Fig. 6.1.

The following system of differential equations is associated with Fig. 6.1:

$$\frac{dP_0(t)}{dt} + n(\lambda_1 + \lambda_h)P_0(t) = 0 \ , \qquad \text{for } n = 1,2,3,\ldots \ , \tag{6.55}$$

$$\frac{dP_1(t)}{dt} + (n-1)(\lambda_1 + \lambda_h)P_1(t) = P_0(t)n\lambda_1 \qquad \text{for } n = 1,2,3\ldots \ , \tag{6.56}$$

$$\frac{dP_2(t)}{dt} + (n-1)(\lambda_1 + \lambda_h)P_2(t) = P_0(t)n\lambda_h \quad \text{for } n = 1,2,3,\ldots \ , \tag{6.57}$$

$$\frac{dP_3(t)}{dt} + (n-2)(\lambda_1 + \lambda_h)P_3(t) = P_1(t)(n-1)\lambda_1 + P_2(t)(n-1)\lambda_1$$

$$\text{for } n = 2,3,4\ldots \ , \tag{6.58}$$

$$\frac{dP_4(t)}{dt} + (n-2)(\lambda_1 + \lambda_h)P_4(t) = P_2(t)(n-1)\lambda_h + P_1(t)(n-1)\lambda_h$$

$$\text{for } n = 2,3,4,\ldots \tag{6.59}$$

.
.
.

$$\frac{dP_{2n-3}(t)}{dt} + (\lambda_1 + \lambda_h)P_{2n-3}(t) = P_{2n-5}(t)2\lambda_1 + P_{2n-4}(t)2\lambda_1$$

$$\text{for } n = 3,4\ldots \ , \tag{6.60}$$

$$\frac{\mathrm{d}P_{2n-2}(t)}{\mathrm{d}t} + (\lambda_1 + \lambda_\mathrm{h})P_{2n-2}(t) = P_{2n-5}(t)2\lambda_\mathrm{h} + P_{2n-4}(t)2\lambda_\mathrm{h}$$

$$\text{for } n = 3,4\ldots\,, \qquad\qquad (6.61)$$

$$\frac{\mathrm{d}P_{2n-1}(t)}{\mathrm{d}t} = P_{2n-3}(t)\lambda_1 + P_{2n-2}(t)\lambda_1 \qquad \text{for } n = 2, 3,\ldots\,, \qquad (6.62)$$

$$\frac{\mathrm{d}P_{2n}(t)}{\mathrm{d}t} = P_{2n-3}(t)\lambda_\mathrm{h} + P_{2n-2}(t)\lambda_\mathrm{h} \qquad \text{for } n = 2, 3,\ldots\,. \qquad (6.63)$$

At $t = 0$, $P_0(0) = 1$, and all other initial condition probabilities are equal to zero.

Laplace transforms of the solution of Eqs. (6.55)–(6.63) are given as follows:

$$P_0(s) = \frac{1}{s + na} \qquad \text{for } n = 1,2,3,\ldots \qquad\qquad (6.64)$$

where $A = \lambda_1 + \lambda_\mathrm{h}$,

$$P_1(s) = \frac{n\lambda_1}{s + (n-1)A} P_0(s) \qquad \text{for } n = 1,2,3,4,\ldots\,, \qquad (6.65)$$

$$P_2(s) = \frac{n\lambda_\mathrm{h}}{s + (n-1)A} P_0(s) \qquad \text{for } n = 1,2,3,4,\ldots\,, \qquad (6.66)$$

$$P_K(s) = \frac{\lambda_1 \displaystyle\prod_{i=0}^{(k-2)/2} (n-i)A^{(K-1)/2}}{\displaystyle\prod_{i=1}^{(k+1)/2} [s + (n-i)A]} P_0(s)$$

$$\text{for } K = 1,3,5,7,\ldots,(2n-3)\,, \quad n = 1,2,3\ldots\,, \qquad (6.67)$$

$$P_K(s) = \frac{\lambda_\mathrm{h} \displaystyle\prod_{i=0}^{(k-2)/2} (n-i)A^{(K-2)/2}}{\displaystyle\prod_{i=1}^{K/2} [s + (n-i)A]} P_0(s)$$

$$\text{for } K = 2,4,6,8,\ldots,(2n-2)\,, \quad n = 1,2,3,\ldots\,, \qquad (6.68)$$

$$P_{2n-1}(s) = \frac{n!\lambda_1 A^{n-1}}{\prod\limits_{i=0}^{n-1}(s+iA)}\, P_0(s) \qquad \text{for } n = 1,2,3,\dots\, , \tag{6.69}$$

$$P_{2n}(s) = \frac{n!\lambda_h A^{n-1}}{\prod\limits_{i=0}^{n-1}(s+iA)}\, P_0(s) \qquad \text{for } n = 1,2,3,\dots\,. \tag{6.70}$$

The Laplace transform of the system reliability is

$$R(s) = \sum_{i=0}^{2n-2} P_i(s) \qquad \text{for } n = 1,2,3,\dots\,. \tag{6.71}$$

The system mean time to failure is given by

$$\text{MTTF} = \lim_{s\to 0} R(s) = \sum_{K=1}^{n} \frac{1}{KA} \qquad \text{for } n = 1,2,3,\dots\,. \tag{6.72}$$

The Laplace transform of the probability that at least one unit is in failed condition due to human error is given by

$$P_h(s) = \sum_{K=1}^{n} P_{2K}(s) \qquad \text{for } n = 1,2,3\dots\,. \tag{6.73}$$

Finally, the Laplace transform of the probability that at least one unit is in failed condition due to hardware failure is

$$P(s) = \sum_{K=1}^{n} P_{2K-1}(s) \qquad \text{for } n = 1,2,3,\dots\,. \tag{6.74}$$

Analysis of a Repairable Two-Unit Parallel System

This section presents a Markov model representing a two-unit active parallel system (i.e., both units operating simultaneously) [4]. The system fails only when both units fail. A unit fails either due to a hardware failure or due to a human error. A failed unit is repaired from each failure state. The repaired unit is put back into operation as soon as the repair is over. The system state-space diagram is shown in Fig. 6.5.

The following symbols are associated with the mathematical model: j is the jth state of the system: $j = 0$ (both units operating normally); $j = 1$ (one unit

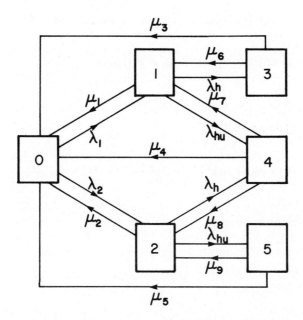

Figure 6.5. State-space diagram for a repairable two-unit parallel system.

failed due to hardware failure, other unit operating); $j = 2$ (one unit failed due to human error, other unit operating); $j = 3$ (both units failed due to hardware failure); $j = 4$ (both units failed — one failed due to human error, the other due to hardware failure); $j = 5$ (both units failed due to human error). $P_j(t)$ is the probability that the system is in state j at time t, for $j = 0,1,2,3,4,5$. λ_h is the constant hardware failure rate of the unit (in Fig. 6.5, $\lambda_1 = 2\lambda_h$). λ_{hu} is the constant human error rate of the unit (in Fig. 6.5, $\lambda_2 = 2\lambda_{hu}$). μ_j is the constant repair rate from system state j: $j = 1,2,3,4,5$ (means the repair rate from states 1, 2, 3, 4, 5 to state 0 respectively); $j = 6,7$ (means the repair rate from states 3 and 4 to state 1, respectively); $j = 8,9$ (means the repair rate from states 4 and 5 to state 2 respectively). S is the Laplace transform variable.

The following assumptions are associated with this model:

1. failures are statistically independent;
2. unit hardware failure and human-error rates are constant;
3. unit and system repair rates are constant;
4. a repaired unit is as good as new;
5. all system units are identical.

The system of differential equations associated with Fig. 6.5 is

$$\frac{dP_0(t)}{dt} + (\lambda_1 + \lambda_2)P_0(t) = P_1(t)\mu_1 + P_2(t)\mu_2 + P_3(t)\mu_3$$
$$+ P_4(t)\mu_4 + P_5(t)\mu_5 \qquad (6.75)$$

where $\lambda_1 = 2\lambda_h$ and $\lambda_2 = 2\lambda_{hu}$,

$$\frac{dP_1(t)}{dt} + (\lambda_h + \lambda_{hu} + \mu_1)P_1(t) = P_0(t)\lambda_1 + P_3(t)\mu_6 + P_4(t)\mu_7 \ , \quad (6.76)$$

$$\frac{dP_2(t)}{dt} + (\lambda_h + \lambda_{hu} + \mu_2)P_2(t) = P_0(t)\lambda_2 + P_4(t)\mu_8 + P_5(t)\mu_9 \ , \quad (6.77)$$

$$\frac{dP_3(t)}{dt} + (\mu_3 + \mu_6)P_3(t) = P_1(t)\lambda_h \ , \qquad (6.78)$$

$$\frac{dP_4(t)}{dt} + (\mu_4 + \mu_7 + \mu_8)P_4(t) = P_1(t)\lambda_{hu} + P_2(t)\lambda_h \ , \qquad (6.79)$$

$$\frac{dP_5(t)}{dt} + (\mu_5 + \mu_9)P_5(t) = P_2(t)\lambda_{hu} \ . \qquad (6.80)$$

At $t = 0$, $P_0(0) = 1$, $P_1(0) = P_2(0) = P_3(0) = P_4(0) = P_5(0) = 0$.

Setting the derivatives of Eqs. (6.75)–(6.80) equal to zero and utilizing the relationship $\sum_{i=0}^{5} P_i = 1$ leads to the following steady-state probability solutions:

$$P_0 = 1 + [A_7(1 + A_1 + A_2\lambda_{hu}) + A_8(1 + A_3 + A_2\lambda_h)]^{-1} \qquad (6.81)$$

where

$$A_1 = \lambda_h/(\mu_3 + \mu_6) \ , \qquad A_2 = 1/(\mu_4 + \mu_7 + \mu_8) \ ,$$

$$A_3 = \lambda_{hu}/(\mu_5 + \mu_9) \ , \qquad A_4 = \lambda_h + \lambda_{hu} + \mu_2 - \lambda_h\mu_8 A_2 - \mu_9 A_3 \ ,$$

$$A_5 = \lambda_h + \lambda_{hu} + \mu_1 - \mu_6 A_1 - \lambda_{hu}\mu_7 A_2 \ , \qquad A_6 = \lambda_h\mu_7 A_2/A_4 \ ,$$

$$A_7 = (\lambda_1 + \lambda_2 A_6)/(A_5 - \lambda_{hu}\mu_8 A_2 A_6) \ ,$$

$$A_8 = (\lambda_2 + \lambda_{hu}\mu_8 A_2 A_7)/A_4 \ ;$$

$$P_1 = P_0 A_7 \; , \tag{6.82}$$

$$P_2 = P_0 A_8 \; , \tag{6.83}$$

$$P_3 = P_0 A_1 A_7 \; , \tag{6.84}$$

$$P_4 = P_0 A_2 (\lambda_{hu} A_7 + \lambda_h A_8) \; , \tag{6.85}$$

$$P_5 = P_0 A_3 A_8 \; . \tag{6.86}$$

The steady-state availability of the system is given by

$$AV_{ss} = P_0 + P_1 + P_2 \; . \tag{6.87}$$

Plots of Eq. (6.87) are shown in Fig. 6.6 for the fixed values of μ_j, $j = 1, 2, \ldots, 9$. As expected, the plots exhibit that the steady-state availability decreases with increasing values of λ_{hu}. Furthermore, the steady-state availability decreases as the value of λ_h increases. The steady-state probability that at least one unit failed due to human error is given by

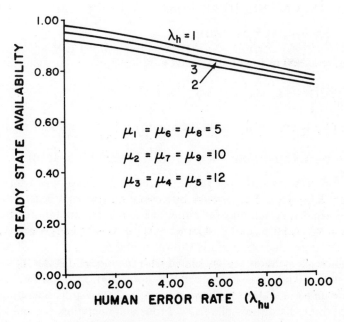

Figure 6.6. Steady-state availability plots.

$$P_{hu} = P_2 + P_4 + P_5 \ . \tag{6.88}$$

The steady-state probability that at least one unit failed due to hardware failure is given by

$$P_h = P_1 + P_3 + P_4 \ . \tag{6.89}$$

The mean time to failure of the system (with repair) is given by

$$\text{MTTF}_R = \lim_{S \to 0} R(S) \ , \tag{6.90}$$

where $R(S)$ is the Laplace transform of the system reliability function.

By setting $\mu_j = 0$ for $j = 3, \ldots, 9$ in Eqs. (6.75)–(6.80), taking the Laplace transforms and solving for $P_0(S)$, $P_1(S)$ and $P_2(S)$, and utilizing the relationship (6.90) yields

$$\text{MTTF}_R = A_{11}/A_{12} \ , \tag{6.91}$$

where

$$A_{11} = (\lambda_h + \lambda_{hu} + \mu_1)(\lambda_2 + \lambda_h + \lambda_{hu} + \mu_2) + \lambda_1(\lambda_h + \lambda_{hu} + \mu_2) \ ,$$

$$A_{12} = (\lambda_h + \lambda_{hu} + \mu_1)[(\lambda_1 + \lambda_2)(\lambda_h + \lambda_{hu} + \mu_2) - \lambda_2\mu_2]$$
$$- \lambda_1\mu_1(\lambda_h + \lambda_{hu} + \mu_2) \ .$$

Setting $\mu_1 = \mu_2 = 0$ in Eq. (6.91) yields

$$\text{MTTF} = (\lambda_1 + \lambda_2 + \lambda_h + \lambda_{hu})/(\lambda_1 + \lambda_2)(\lambda_h + \lambda_{hu}) \ , \tag{6.92}$$

where MTTF is the mean time to failure of the system without repair.

Analysis of a Repairable Two-out-of-Three Unit System

This section presents a Markov model representing a two-out-of-three unit active parallel system. The system is successful only when at least two units operate normally. A unit may fail either due to a hardware failure or due to a human error. A failed unit is repaired back to its normal operation mode. The system state-space diagram is shown in Fig. 6.7.

The following symbols are associated with the model: j is the jth state of the system: $j = 0$ (all three units operating normally); $j = 1$ (two units operating, one unit failed due to human error); $j = 2$ (two units operating, one unit failed due to hardware failure); $j = 3$ (one unit operating, two units failed

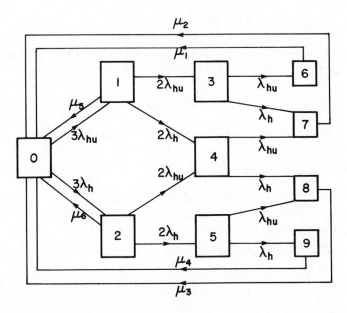

Figure 6.7. State-space diagram for a repairable two-out-of-three unit system.

due to human error); $j = 4$ (one unit operating, two units failed — one unit failed due to human error and the other due to hardware failure); $j = 5$ (one unit operating, two units failed due to hardware failure); $j = 6$ (three units failed due to human error); $j = 7$ (three units failed — two units failed due to human error and the other unit due to hardware failure); $j = 8$ (three units failed — two units failed due to hardware failure and the other unit due to human error); $j = 9$ (three units failed due to hardware failure). $P_j(t)$ is the probability that the system is in state j at time t, for $j = 0,1,2,\ldots,9$. λ_h is the constant hardware failure rate of the unit. λ_{hu} is the constant human error rate of the unit. μ_j is the jth constant repair rate: $j = 1$ (state 6 to state 0), $j = 2$ (state 7 to state 0), $j = 3$ (state 8 to state 0), $j = 4$ (state 9 to state 0), $j = 5$ (state 1 to state 0), $j = 6$ (state 2 to state 0). S is the Laplace transform variable.

The following assumptions are associated with this Markov model:

1. all system units are identical;
2. unit and system repair rates are constant;
3. failures are statistically independent;
4. a repaired unit is as good as new;
5. unit hardware failure and human-error rates are constant.

The system of differential equations associated with Figure 6.7 is

$$\frac{\mathrm{d}P_0(t)}{\mathrm{d}t} + 3(\lambda_{hu} + \lambda_h)P_0(t) = P_1(t)\mu_5 + P_2(t)\mu_6 + P_6(t)\mu_1 + P_7(t)\mu_2$$

$$+ P_8(t)\mu_3 + P_9(t)\mu_4 \ , \tag{6.93}$$

$$\frac{\mathrm{d}P_1(t)}{\mathrm{d}t} + (2\lambda_{hu} + 2\lambda_h + \mu_5)P_1(t) = P_0(t)3\lambda_{hu} \ , \tag{6.94}$$

$$\frac{\mathrm{d}P_2(t)}{\mathrm{d}t} + (2\lambda_{hu} + 2\lambda_h + \mu_6)P_2(t) = P_0(t)3\lambda_h \ , \tag{6.95}$$

$$\frac{\mathrm{d}P_3(t)}{\mathrm{d}t} + (\lambda_{hu} + \lambda_h)P_3(t) = P_1(t)2\lambda_{hu} \ , \tag{6.96}$$

$$\frac{\mathrm{d}P_4(t)}{\mathrm{d}t} + (\lambda_{hu} + \lambda_h)P_4(t) = P_1(t)2\lambda_h + P_2(t)2\lambda_{hu} \ , \tag{6.97}$$

$$\frac{\mathrm{d}P_5(t)}{\mathrm{d}t} + (\lambda_{hu} + \lambda_h)P_5(t) = P_2(t)2\lambda_h \ , \tag{6.98}$$

$$\frac{\mathrm{d}P_6(t)}{\mathrm{d}t} + \mu_1 P_6(t) = P_3(t)\lambda_{hu} \ , \tag{6.99}$$

$$\frac{\mathrm{d}P_7(t)}{\mathrm{d}t} + \mu_2 P_7(t) = P_3(t)\lambda_h + P_4(t)\lambda_{hu} \ , \tag{6.100}$$

$$\frac{\mathrm{d}P_8(t)}{\mathrm{d}t} + \mu_3 P_8(t) = P_4(t)\lambda_h + P_5(t)\lambda_{hu} \ , \tag{6.101}$$

$$\frac{\mathrm{d}P_9(t)}{\mathrm{d}t} + \mu_4 P_9(t) = P_5(t)\lambda_h \ . \tag{6.102}$$

At $t = 0$, $P_0(0) = 1$, and all other initial-condition probabilities are equal to zero.

Setting the derivatives of Eqs. (6.93)–(6.102) equal to zero and using the relationship $\sum_{i=0}^{9} P_i = 1$ leads to the following steady-state probability solutions:

$$P_0 = B_2/(B_1 + B_2) \ , \tag{6.103}$$

where

$$B_1 = 2\lambda_h\mu_1\mu_2\mu_3(\mu_4 B + 2\lambda_4\mu_4 + 2\lambda_h^2)(2B + \mu_5)$$
$$+ 3\lambda_{hu}\mu_2\mu_3\mu_4(\mu_1 B + 2\lambda_{hu}\mu_1 + 2\lambda_{hu}^2)(2B + \mu_6)$$
$$+ 6\lambda_{hu}\lambda_h\mu_1\mu_2\mu_3\mu_4(4B + \mu_4 + \mu_6)$$
$$+ 6\lambda_{hu}^2\lambda_h\mu_1\mu_3\mu_4(6B + \mu_5 + 2\mu_6)$$
$$+ 6\lambda_h^2\lambda_{hu}\mu_1\mu_2\mu_4(6B + 2\mu_5 + \mu_6) \ ;$$

$$B_2 = \mu_1\mu_2\mu_3\mu_4 B(2B + \mu_5)(2B + \mu_6) \ , \qquad B = \lambda_{hu} + \lambda_h \ ;$$

$$P_1 = P_0 3\lambda_{hu}/(2B + \mu_5) \ , \tag{6.104}$$

$$P_2 = P_0 3\lambda_h/(2B + \mu_6) \ , \tag{6.105}$$

$$P_3 = P_0 6\lambda_{hu}^2/B(2B + \mu_5) \ , \tag{6.106}$$

$$P_4 = P_0 6\lambda_{hu}\lambda_h(4B + \mu_5 + \mu_6)/B(2B + \mu_5)(2B + \mu_6) \ , \tag{6.107}$$

$$P_5 = P_0 6\lambda_h^2/B(2B + \mu_6) \ , \tag{6.108}$$

$$P_6 = P_0 6\lambda_{hu}^3/\mu_1 B(2B + \mu_5) \ , \tag{6.109}$$

$$P_7 = P_0 6\lambda_{hu}^2\lambda_h(6B + \mu_5 + 2\mu_6)/\mu_2 B(2B + \mu_5)(2B + \mu_6) \ , \tag{6.110}$$

$$P_8 = P_0 6\lambda_h^2\lambda_{hu}(6B + 2\mu_5 + \mu_6)/\mu_3 B(2B + \mu_5)(2B + \mu_6) \ , \tag{6.111}$$

$$P_9 = P_0 6\lambda_h^3/\mu_4 B(2B + \mu_6) \ . \tag{6.112}$$

The steady-state availability of the system is given by

$$AV_{ss} = P_0 + P_1 + P_2 \ . \tag{6.113}$$

The steady-state probability that at least one unit failed due to human error is given by

$$P_{hu} = P_1 + P_3 + P_4 + P_6 + P_7 + P_8 \ . \tag{6.114}$$

The steady-state probability that at least one unit failed due to hardware failure is given by

$$P_h = P_2 + P_4 + P_5 + P_7 + P_8 + P_9 \ . \tag{6.115}$$

Setting $\mu_1 = \mu_2 = \mu_3 = \mu_4 = 0$ in Eqs. (6.93)–(6.102), taking the Laplace transforms, solving for $P_0(S)$, $P_1(S)$ and $P_2(S)$ and utilizing relationship (6.90) leads to

$$\mathrm{MTTF}_R = \lim_{s \to 0} R(S) = \lim_{s \to 0} \{P_0(S) + P_1(S) + P_2(S)\}$$

$$= \frac{(2B + \mu_5)(2B + \mu_6 + 3\lambda_h) + 3\lambda_{hu}(2B + \mu_6)}{3(2B + \mu_5)[B(2B + \mu_6) - \lambda_h\mu_6] - 3\lambda_{hu}\mu_5(2B + \mu_6)} \quad , \quad (6.116)$$

where MTTF_R is the mean time to failure of the system with repair. By setting $\mu_5 = \mu_6 = 0$ in Eq. (6.116) we get

$$\mathrm{MTTF} = 5/6B \ , \tag{6.117}$$

where MTTF is the mean time to failure of the system without repair.

The plots of Eqs. (6.116) and (6.117) are shown in Fig. 6.8 for fixed values of λ_h, μ_5 and μ_6. These plots indicate that mean time to failure decreases with the increasing values of λ_{hu}. Furthermore, the mean time to failure with repair is higher than the mean time to failure without repair.

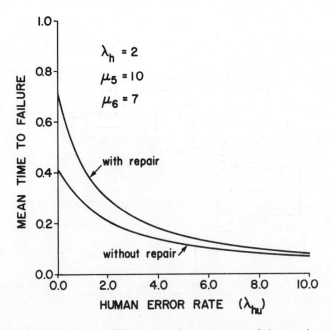

Figure 6.8. Mean time to failure plots for a two-out-of-three unit system.

Analysis of a Two-Identical-Unit Standby System

This section presents a Markov model representing a two-identical-unit standby system. Initially, at $t = 0$ one unit starts operating and the other is on standby. Failure rate of the standby unit is zero. The operating unit can fail due to a hardware failure or a human error. As soon as the operating unit fails, the standby unit is switched into operation. A failed unit is repaired back to its normal operation mode. The state-space diagram of this Markov model is shown in Fig. 6.9.

The notations used to develop the state probability equations for this model are defined as follows: j is the jth state of the system: $j = 0$ (one unit operating, other unit on standby); $j = 1$ (one unit failed due to hardware failure, the other unit operating); $j = 2$ (one unit failed due to human error, the other unit operating); $j = 3$ (both units failed due to hardware failure), $j = 4$ (both units failed—one due to human error, the other due to hardware failure); $j = 5$ (both units failed due to human error). $P_j(t)$ is the probability that the system is in state j at time t, for $j = 0,1,2,3,4,5$. λ_h is the constant hardware failure rate of the unit. λ_{hu} is the constant human error rate of the unit. μ_j is the jth constant repair rate: $j = 1$ (state 1 to state 0), $j = 2$ (state 2 to state 0), $j = 3$ (state 3 to state 0), $j = 4$ (state 4 to state 0), $j = 5$ (state 5 to state 0), $j = 6$ (state 3 to state 1), $j = 7$ (state 4 to state 1), $j = 8$ (state 4 to state 2), $j = 0$ (state 5 to state 2). S is the Laplace transform variable.

The following assumptions are associated with this Markov model:

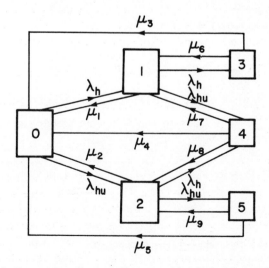

Figure 6.9. State-space diagram for a two-identical-unit standby system.

1. a repaired unit is as good as new;
2. failures are statistically independent;
3. all system units are identical;
4. unit and system repair rates are constant;
5. unit hardware failure and human-error rates are constant;
6. the switchover mechanism is perfect.

The following differential equations are associated with Fig. 6.9:

$$\frac{dP_0(t)}{dt} + (\lambda_h + \lambda_{hu})P_0(t) = P_1(t)\mu_1 + P_2(t)\mu_2 + P_3(t)\mu_3$$

$$+ P_4(t)\mu_4 + P_5(t)\mu_5 \ , \tag{6.118}$$

$$\frac{dP_1(t)}{dt} + (\lambda_h + \lambda_{hu} + \mu_1)P_1(t) = P_0(t)\lambda_h + P_3(t)\mu_6 + P_4(t)\mu_7 \ , \tag{6.119}$$

$$\frac{dP_2(t)}{dt} + (\lambda_h + \lambda_{hu} + \mu_2)P_2(t) = P_0(t)\lambda_{hu} + P_4(t)\mu_8 + P_5(t)\mu_9 \ , \tag{6.120}$$

$$\frac{dP_3(t)}{dt} + (\mu_3 + \mu_6)P_3(t) = P_1(t)\lambda_h \ , \tag{6.121}$$

$$\frac{dP_4(t)}{dt} + (\mu_4 + \mu_7 + \mu_8)P_4(t) = P_1(t)\lambda_{hu} + P_2(t)\lambda_h \ , \tag{6.122}$$

$$\frac{dP_5(t)}{dt} + (\mu_5 + \mu_9)P_5(t) = P_2(t)\lambda_{hu} \ . \tag{6.123}$$

At $t = 0$, $P_0(0) = 1$, and all other initial condition probabilities are equal to zero.

Setting the derivatives of Eqs. (6.118)–(6.123) equal to zero and utilizing the relationship $\sum_{i=0}^{5} P_i = 1$ leads to the following steady-state probability solutions:

$$P_0 = [1 + A_7(1 + A_1 + A_2\lambda_{hu}) + A_8(1 + A_3 + A_2\lambda_h)]^{-1} \ , \tag{6.124}$$

where

$$A_1 = \lambda_h/(\mu_3 + \mu_6) \ , \qquad A_2 = 1/(\mu_4 + \mu_7 + \mu_8) \ ,$$

$$A_3 = \lambda_{hu}/(\mu_5 + \mu_9) \ , \qquad A_4 = \lambda_h + \lambda_{hu} + \mu_2 - \lambda_h\mu_8 A_2 - \mu_9 A_3 \ ,$$

$$A_5 = \lambda_h + \lambda_{hu} + \mu_1 - \mu_6 A_1 - \lambda_{hu}\mu_7 A_2 \ , \qquad A_6 = \lambda_h \mu_7 A_2 / A_4 \ ,$$

$$A_7 = (\lambda_h + \lambda_{hu} A_6)/(A_5 - \lambda_{hu}\mu_8 A_2 A_6) \ , \qquad A_8 = \lambda_{hu}(1 + \mu_8 A_2 A_7)/A_4 \ ;$$

$$P_1 = P_0 A_7 \ , \tag{6.125}$$

$$P_2 = P_0 A_8 \ , \tag{6.126}$$

$$P_3 = P_0 A_1 A_7 \ , \tag{6.127}$$

$$P_4 = P_0 A_2 (\lambda_{hu} A_7 + \lambda_h A_8) \ , \tag{6.128}$$

$$P_5 = P_0 A_3 A_8 \ . \tag{6.129}$$

The steady-state availability of the system is given by

$$AV_{ss} = P_0 + P_1 + P_2 \ . \tag{6.130}$$

For fixed values of μ_j $(j = 1, 2, \ldots, 9)$ and λ_h, the plots of Eq. (6.130) are shown in Fig. 6.10. These plots exhibit the effect of varying λ_{hu} from 0 to

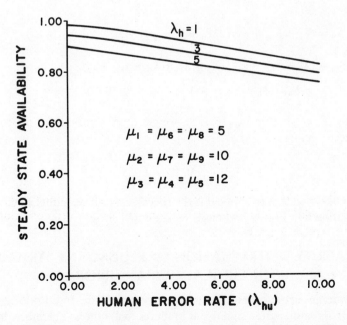

Figure 6.10. Steady-state availability plots for a two-identical-unit standby system.

10 for different values of λ_h (= 1,3,5) on steady-state availability. As expected, the steady-state availability decreases with increasing values of λ_{hu} and λ_h.

The steady-state probability that at least one unit failed due to human error is given by

$$P_{hu} = P_2 + P_4 + P_5 \ . \tag{6.131}$$

The steady-state probability that at least one unit failed due to hardware failure is given by

$$P_h = P_1 + P_3 + P_4 \ . \tag{6.132}$$

Setting $\mu_j = 0$ for $j = 3, \ldots 9$ in Eqs. (6.118)–(6.123), taking the Laplace transforms, solving for $P_0(S)$, $P_1(S)$ and $P_2(S)$ and utilizing relationship (6.90) leads to the mean time to failure with repair:

$$\text{MTTF}_R = \lim_{S \to 0} R(S) = \lim_{S \to 0} \{P_0(S) + P_1(S) + P_2(S)\}$$

$$= A_9/A_{10} \ , \tag{6.133}$$

where

$$A_9 = (\lambda_h + \lambda_{hu} + \mu_1)(2\lambda_{hu} + \lambda_h + \mu_2) + \lambda_h(\lambda_h + \lambda_{hu} + \mu_2) \ ,$$

$$A_{10} = (\lambda_h + \lambda_{hu} + \mu_1)[(\lambda_h + \lambda_{hu})(\lambda_h + \lambda_{hu} + \mu_2) - \lambda_{hu}\mu_2]$$
$$- \lambda_h\mu_1(\lambda_h + \lambda_{hu} + \mu_2) \ .$$

By setting $\mu_1 = \mu_2 = 0$ in Eq. (6.133), we get the system mean time to failure without repair:

$$\text{MTTF} = 2/(\lambda_h + \lambda_{hu}) \ . \tag{6.134}$$

It is to be noted that this model is the special case of the model state-space diagram shown in Fig. 6.5 when $\lambda_1 = \lambda_h$ and $\lambda_2 = \lambda_{hu}$.

RELIABILITY DETERMINATION OF REDUNDANT SYSTEMS WITH CRITICAL HUMAN ERROR

Not all human errors necessarily cause system failures. In addition, some human errors cause more "significant" failures than others. Therefore, in this section the errors causing significant failures are called critical human errors.

Once a critical human error takes place the entire system fails. Thus it is assumed that for each system unit the failures due to critical human errors are separated from the hardware failures. An example of a critical human error is a fire caused by humans in a room where redundant units are located. In this case the entire redundant system will fail, irrespective of whether one or more units were operating.

This section presents a number of Markov models to evaluate the reliability of various kinds of redundant systems with critical human errors [4–6].

Model I

This model represents a nonrepairable two-identical-unit parallel system. The system transition diagram is shown in Fig. 6.11. The system fails due to critical human errors or hardware failures. This model separates only the critical human errors from the hardware failures; in other words, only those human errors due to which both units fail (i.e., when both units were operating normally) or would have failed (i.e., when only one unit was operating normally). More clearly, due to a human action the system fails when both units are functioning normally. Furthermore, when only one unit is operating normally, due to the same human action (i.e., that action which caused both units to fail simultaneously) the operating unit failed. In other words, if both of the units had been operating normally instead of only one, the entire system would have failed due to the same action. For example, if there had been a fire in a room due to a human error where both units are located, the entire system (i.e., two-unit system) would fail irrespective of whether one or two units are operating successfully.

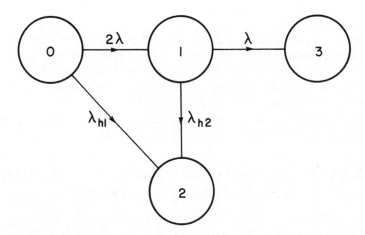

Figure 6.11. System transition diagram for a two-unit parallel system.

The following notation is used with this model. 0 in the circle of Fig. 6.11 denotes that both the units in the system are in the operating state. 1 in the circle of Fig. 6.11 denotes that only one of the units is operating. 2 in the circle of Fig. 6.11 denotes that the system is in the failed state due to critical human error. 3 in the circle of Fig. 6.11 denotes that the system is in the failed state due to hardware failures plus noncritical human errors. λ is the unit constant failure rate (this also includes noncritical human errors). λ_{h1} is the constant critical human error rate when two units are in the operating state. λ_{h2} is the constant critical human error rate when only one unit is in the operating state. $P_i(t)$ is the probability that the system is in ith state at time t, for $i = 0,1,2,3$. $R(t)$ is the system reliability at time t. MTTF is the system mean time to failure. S is the Laplace transform variable.

The following assumptions are associated with this model.

1. Critical human errors may occur when either both system units are good or when one system unit is good.
2. Both units are identical and active.
3. Failures are statistically independent.
4. Failure and critical human error rates are constant.
5. The entire system fails due to critical human errors.

The system of first-order differential equations associated with Fig. 6.11 is

$$P_0'(t) = -(2\lambda + \lambda_{h1})P_0(t) \ , \tag{6.135}$$

$$P_1'(t) = -(\lambda + \lambda_{h2})P_1(t) + 2\lambda P_0(t) \ , \tag{6.136}$$

$$P_2'(t) = \lambda_{h1}P_0(t) + \lambda_{h2}P_1(t) \ , \tag{6.137}$$

$$P_3'(t) = \lambda P_1(t) \ , \tag{6.138}$$

where the prime denotes the differentiation with respect to time t. At $t = 0$, $P_0(t) = 1$, and other initial condition probabilities are equal to zero.

Solving the set of Eqs. (6.135)–(6.138) yields the resulting state probabilities in terms of Laplace transforms as follows:

$$P_0(s) = \frac{1}{s + a_1} \ , \tag{6.139}$$

$$P_1(s) = \frac{2\lambda}{(s + a_1)(s + a_2)} \ , \tag{6.140}$$

$$P_2(s) = \frac{[2\lambda\lambda_{h2} + \lambda_{h1}(s + a_2)]}{s(s + a_1)(s + a_2)} \ , \tag{6.141}$$

and

$$P_3(s) = \frac{2\lambda^2}{s(s + a_1)(s + a_2)} \ , \tag{6.142}$$

where

$$a_1 = 2\lambda + \lambda_{h1} \ ,$$

$$a_2 = \lambda + \lambda_{h2} \ .$$

The state probabilities are obtained by taking the inverse Laplace transforms of Eqs. (6.139)-(6.142) and are given by

$$P_0(t) = e^{-a_1 t} \ , \tag{6.143}$$

$$P_1(t) = A_1(e^{-a_1 t} - e^{-a_2 t}) \ , \tag{6.144}$$

$$P_2(t) = A_2 - A_3 e^{-a_1 t} - A_4 e^{-a_2 t} \ , \tag{6.145}$$

$$P_3(t) = A_5 - A_6 e^{-a_1 t} - A_7 e^{-a_2 t} \ , \tag{6.146}$$

where

$$A_1 \equiv \frac{2\lambda}{a_2 - a_1} \ ,$$

$$A_2 \equiv \frac{2\lambda\lambda_{h2} + \lambda_{h1} a_2}{a_1 a_2} \ ,$$

$$A_3 \equiv \frac{2\lambda\lambda_{h2} + \lambda_{h1}(a_2 - a_1)}{a_1(a_2 - a_1)} \ ,$$

$$A_4 \equiv \frac{2\lambda\lambda_{h2}}{a_2(a_1 - a_2)} \ ,$$

$$A_5 = \frac{2\lambda^2}{a_1 a_2} \ ,$$

$$A_6 = \frac{2\lambda^2}{a_1(a_2 - a_1)} \ ,$$

$$A_7 = \frac{2\lambda^2}{a_2(a_1 - a_2)} \ .$$

The system reliability expression in terms of Laplace transforms is obtained by adding Eqs. (6.139) and (6.140):

$$R(s) = P_0(s) + P_1(s)$$

or

$$R(s) = \frac{s + a_3}{(s + a_1)(s + a_2)} , \qquad (6.147)$$

where $a_3 = a_2 + 2\lambda$. The system MTTF is given by

$$MTTF = \lim_{s \to 0} R(s)$$

$$= \lim_{s \to 0} \frac{s + a_3}{(s + a_1)(s + a_2)} ,$$

or

$$MTTF = \frac{a_3}{a_1 a_2} = \frac{3\lambda + \lambda_{h2}}{(2\lambda + \lambda_{h1})(\lambda + \lambda_{h2})} . \qquad (6.148)$$

From Eq. (6.147) the system reliability is given by

$$R(t) = \mathcal{L}^{-1} \left\{ \frac{s + a_3}{(s + a_1)(s + a_2)} \right\}$$

or

$$R(t) = (1 + A_1) e^{-a_1 t} - A_1 e^{-a_2 t} . \qquad (6.149)$$

EXAMPLE 6.4

A system is composed of two identical and independent units in parallel. Each unit constant failure rate is 0.01 failure/hr. The value of the critical human error rate when both units are operating is 0.005 error/hr. However, when only one unit is operating the value of the critical human error rate is 0.002 error/hr. Compute the parallel system mean time to failure.

In this example, the following values are given for λ, λ_{h1} and λ_{h2}: $\lambda = 0.01$ failure/hr, $\lambda_{h1} = 0.005$ error/hr, and $\lambda_{h2} = 0.002$ error/hr.

By utilizing the data given above and inserting into Eq. (6.148), we get

$$MTTF = \frac{3\lambda + \lambda_{h2}}{(2\lambda + \lambda_{h1})(\lambda + \lambda_{h2})} = 106.67 \text{ hr} .$$

It is interesting to note from the above result that the system MTTF has increased only 6.67 hr, even with an additional (redundant) unit. The unit MTTF is 100 hr.

Model II

This Markov model is concerned with a two-identical-unit parallel system subject to hardware failure and critical human error [4]. In any operating state, the entire system fails due to the occurrence of critical human error. Fire due to human error in a room containing a parallel system is a typical example. The failed unit is repaired back to its normal operation mode. The model state-space diagram is shown in Fig. 6.12.

The following symbols were used to develop the system of differential equations for the model. j is the jth state of the system: $j = 0$ (both units operating); $j = 1$ (one unit failed due to hardware failure, the other unit operating); $j = 2$ (both units failed due to hardware failure); $j = 3$ (system failed due to a critical human error). $P_j(t)$ is the probability that the system is in state j at time t, for $j = 0, 1, 2, 3$. λ is the constant hardware failure rate of the unit. λ_{h1} is the constant critical human error rate from state 0 to state 3. λ_{h2} is the constant critical human error rate from state 1 to state 3. μ_j is the jth constant repair rate: $j = 1$ (state 2 to state 0), $j = 2$ (state 1 to state 0), $j = 3$ (state 3 to to state 0), $j = 4$ (state 2 to state 1). S is the Laplace transform variable.

The following assumptions are concerned with this model.

1. The repaired unit or the system is as good as new.
2. Both system units are identical and active.
3. The unit failure rate is constant.
4. Critical human error rates are constant.

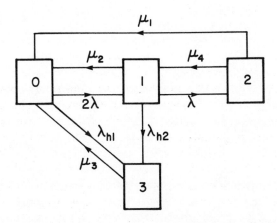

Figure 6.12. State-space diagram for a repairable two-identical-unit parallel system.

5. Failures are statistically independent.
6. Repair rates are constant.

The following differential equations are associated with Fig. 6.12:

$$\frac{dP_0(t)}{dt} + (2\lambda + \lambda_{h1})P_0(t) = P_1(t)\mu_2 + P_2(t)\mu_1 + P_3(t)\mu_3 \ , \qquad (6.150)$$

$$\frac{dP_1(t)}{dt} + (\lambda + \lambda_{h2} + \mu_2)P_1(t) = P_0(t)2\lambda + P_2(t)\mu_4 \ , \qquad (6.151)$$

$$\frac{dP_2(t)}{dt} + (\mu_1 + \mu_4)P_2(t) = P_1(t)\lambda \ , \qquad (6.152)$$

$$\frac{dP_3(t)}{dt} + \mu_3 P_3(t) = P_0(t)\lambda_{h1} + P_1(t)\lambda_{h2} \ . \qquad (6.153)$$

At $t = 0$, $P_0(0) = 1$, $P_1(0) = P_2(0) = P_3(0) = 0$.

The following steady-state probability equations were obtained by setting the derivatives of Eqs. (6.150)–(6.153) equal to zero and utilizing the relationship $P_0 + P_1 + P_2 + P_3 = 1$:

$$P_0 = \left[1 + C_1 + 2\lambda^2 C + \frac{1}{\mu_3}(\lambda_{h1} + C_1 \lambda_{h2}) \right]^{-1} , \qquad (6.154)$$

where

$$C = [(\mu_1 + \mu_4)(\lambda + \lambda_{h2} + \mu_2) - \lambda\mu_4]^{-1} \ ,$$

$$C_1 = 2\lambda(1 + \lambda\mu_4 C)/(\lambda + \lambda_{h2} + \mu_2) \ ,$$

$$P_1 = P_0 C_1 \ , \qquad (6.155)$$

$$P_2 = P_0 2\lambda^2 C \ , \qquad (6.156)$$

$$P_3 = P_0 \frac{1}{\mu_3}(\lambda_{h1} + C_1 \lambda_{h2}) \ . \qquad (6.157)$$

The steady-state availability of the system is given by

$$AV_{ss} = P_0 + P_1 \ . \qquad (6.158)$$

The probability of system failure due to critical human error is given by

$$P_{\text{che}} = P_3 \ . \tag{6.159}$$

The probability that at least one unit failed due to hardware failure is given by

$$P_{\text{h}} = P_1 + P_2 \ . \tag{6.160}$$

Setting $\mu_1 = \mu_3 = \mu_4 = 0$ in Eqs. (6.150)–(6.153), taking the Laplace transforms, solving for $P_0(S)$ and $P_1(S)$ and utilizing the relationship (6.90) leads to

$$\text{MTTF}_R = \lim_{S \to 0} R(S) = \lim_{S \to 0} \{P_0(S) + P_1(S)\}$$

$$= (3\lambda + \lambda_{h2} + \mu_2)/[(2\lambda + \lambda_{h1})(\lambda + \lambda_{h2} + \mu_2) - 2\lambda\mu_2] \ , \tag{6.161}$$

where MTTF_R is the mean time to failure with repair.

Setting $\mu_2 = 0$ in Eq. (6.161) yields the mean time to failure without repair:

$$\text{MTTF} = (3\lambda + \lambda_{h2})/(2\lambda + \lambda_{h1})(\lambda + \lambda_{h2}) \ . \tag{6.162}$$

EXAMPLE 6.5

The following data are specified for the parameters of Eq. (6.161): $\lambda = 0.02$ failure/hr, $\lambda_{h1} = 0.006$ error/hr, $\lambda_{h2} = 0.001$ error/hr, and $\mu_2 = 0.04$ repair/hr.

Calculate the value of the model II mean time to failure with repair.

With the aid of given data and Eq. (6.161), the resulting value of the mean time to failure with repair is

$$
\begin{aligned}
\text{MTTF} &= \frac{(3\lambda + \lambda_{h2} + \mu_2)}{[(2\lambda + \lambda_{h1})(\lambda + \lambda_{h2} + \mu_2) - 2\lambda\mu_2]} \\[2mm]
&= \frac{3(0.02) + 0.001 + 0.04}{[(0.04 + 0.006)(0.02 + 0.001 + 0.04) - 2(0.02)(0.04)]} \\[2mm]
&= 83.75 \text{ hr} \ .
\end{aligned}
$$

Thus the system mean time to failure with repair is 83.75 hr.

Model III

This model represents a two-identical-unit active parallel system with hardware failures and critical and noncritical human errors [4]. Both units operate simultaneously. A unit fails either due to a noncritical human error or due to a hardware failure. The occurrence of critical human error at any system operating state causes the total system failure. A failed unit is repaired back to its normal operation mode. The state-space diagram is shown in Fig. 6.13.

The following symbols are associated with the mathematical model. j is the jth state of the system: $j = 0$ (both units operating); $j = 1$ (one unit failed due to hardware failure, the other unit operating); $j = 2$ (one unit failed due to noncritical human error, the other unit operating); $j = 3$ (both units failed due to hardware failure); $j = 4$ (both units failed — one unit failed due to noncritical human error, the other unit due to hardware failure); $j = 5$ (both units failed due to noncritical human error); $j = 6$ (system failed due to critical human error). $P_j(t)$ is the probability that the system is in state j at time t, for $j = 0, 1, 2, \ldots, 6$. λ_h is the constant hardware failure rate of a unit. λ_{hu} is the constant noncritical human-error rate of a unit. λ_c is the constant critical human-error rate from states 1 and 2 to state 6. λ_{c1} is the constant critical human error rate from state 0 to state 6. μ_j is the jth constant repair rate: $j = 1$ (state 1 to state 0), $j = 2$ (state 2 to state 0), $j = 3$ (state 3 to state 0), $j = 4$ (state 4 to state 0), $j = 5$ (state 5 to state 0), $j = 6$ (state 6 to state

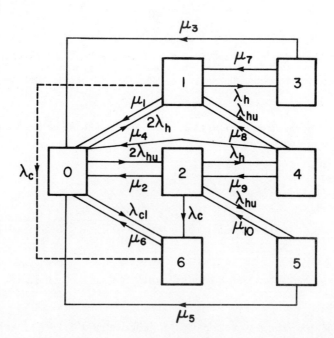

Figure 6.13. State-space diagram for a parallel system with critical and noncritical human errors.

0), $j = 7$ (state 3 to state 1), $j = 8$ (state 4 to state 1), $j = 9$ (state 4 to state 2), $j = 10$ (state 5 to state 2). S is the Laplace transform variable.

The following assumptions are associated with this model.

1. Both system units are identical and active.
2. The repaired unit or the system is as good as new.
3. Unit failure and noncritical error rates are constant.
4. Critical human-error rates are constant.
5. Failures are statistically independent.
6. Repair rates are constant.

The following system of differential equations are associated with Figure 6.13:

$$\frac{dP_0(t)}{dt} + (2\lambda_h + 2\lambda_{hu} + \lambda_{c1})P_0(t) = P_1(t)\mu_1 + P_2(t)\mu_2 + P_3(t)\mu_3 + P_4(t)\,\mu_4$$
$$+ P_5(t)\mu_5 + P_6(t)\mu_6 , \qquad (6.163)$$

$$\frac{dP_1(t)}{dt} + (\lambda_h + \lambda_{hu} + \lambda_c + \mu_1)P_1(t) = P_0(t)2\lambda_h + P_3(t)\mu_7 + P_4(t)\mu_8 ,$$
$$(6.164)$$

$$\frac{dP_2(t)}{dt} + (\lambda_h + \lambda_{hu} + \lambda_c + \mu_2)P_2(t) = P_0(t)2\lambda_{hu} + P_4(t)\mu_9 + P_5(t)\mu_{10} ,$$
$$(6.165)$$

$$\frac{dP_3(t)}{dt} + (\mu_3 + \mu_7)P_3(t) = P_1(t)\lambda_h , \qquad (6.166)$$

$$\frac{dP_4(t)}{dt} + (\mu_4 + \mu_8 + \mu_9)P_4(t) = P_1(t)\lambda_{hu} + P_2(t)\lambda_h , \qquad (6.167)$$

$$\frac{dP_5(t)}{dt} + (\mu_5 + \mu_{10})P_5(t) = P_2(t)\lambda_{hu} , \qquad (6.168)$$

$$\frac{dP_6(t)}{dt} + \mu_6 P_6(t) = P_0(t)\lambda_{c1} + P_1(t)\lambda_c + P_2(t)\lambda_c . \qquad (6.169)$$

At $t = 0$, $P_0(0) = 1$, and all other initial condition probabilities are equal to zero.

Setting the derivatives of Eqs. (6.163)–(6.169) equal to zero and utilizing the relationship $\sum\limits_{i=0}^{6} P_i = 1$ leads to the following steady-state probability solutions:

$$P_0 = \left[1 + D_8(1 + D_3 + \lambda_{hu} D_4) + D_7(1 + D_5 + \lambda_h D_4) \right.$$
$$\left. + \frac{1}{\mu_6}(\lambda_{c1} + \lambda_c D_8 + \lambda_c D_7) \right]^{-1} \tag{6.170}$$

where

$$D_1 = [\lambda_h + \lambda_{hu} + \lambda_c + \mu_1]^{-1} , \qquad D_2 = [\lambda_h + \lambda_{hu} + \lambda_c + \mu_2]^{-1} ,$$

$$D_3 = \lambda_h[\mu_3 + \mu_7]^{-1} , \qquad D_4 = [\mu_4 + \mu_8 + \mu_9]^{-1} ,$$

$$D_5 = \lambda_{hu}[\mu_5 + \mu_{10}]^{-1} , \qquad D_6 = [1 - \mu_7 D_1 D_3 - \lambda_{hu}\mu_8 D_1 D_4]^{-1} ,$$

$$D_7 = 2\lambda_{hu} D_2(1 + \lambda_h\mu_9 D_1 D_4 D_6)[1 - \lambda_h\mu_9 D_2 D_4 - \mu_{10} D_2 D_5$$
$$- \lambda_h\lambda_{hu}\mu_8\mu_9 D_1 D_2 D_4^2 D_6]^{-1} ,$$

$$D_8 = \lambda_h D_1 D_6(2 + \mu_8 D_4 D_7) ;$$

$$P_1 = P_0 D_8 , \tag{6.171}$$

$$P_2 = P_0 D_7 , \tag{6.172}$$

$$P_3 = P_0 D_3 D_8 , \tag{6.173}$$

$$P_4 = P_0(\lambda_{hu} D_8 + \lambda_h D_7)D_4 , \tag{6.174}$$

$$P_5 = P_0 D_5 D_7 , \tag{6.175}$$

$$P_6 = P_0 \frac{1}{\mu_6}(\lambda_{c1} + \lambda_c D_8 + \lambda_c D_7) . \tag{6.176}$$

The steady-state availability of the system is given by

$$AV_{ss} = P_0 + P_1 + P_2 . \tag{6.177}$$

The probability of system failure due to critical human error is given by

$$P_{che} = P_6 . \tag{6.178}$$

The probability that at least one unit failed due to noncritical human error is given by

$$P'_{hu} = P_2 + P_4 + P_5 \ . \tag{6.179}$$

The probability that at least one unit failed due to hardware failure is given by

$$P_h = P_1 + P_3 + P_4 \ . \tag{6.180}$$

Setting $\mu_j = 0$ for $j = 3, \dots, 9$ in Eqs. (6.163)–(6.169), taking the Laplace transforms, solving for $P_0(S)$, $P_1(S)$, and $P_2(S)$, and utilizing relationship (6.90) leads to

$$\mathrm{MTTF_R} = \lim_{S \to 0} R(S) = \lim_{S \to 0} \{ P_0(S) + P_1(S) + P_2(S) \}$$

$$= D_9 / D_{10} \ , \tag{6.181}$$

where

$$D_9 = (\lambda_h + \lambda_{hu} + \lambda_c + \mu_1)(3\lambda_{hu} + \lambda_h + \lambda_c + \mu_2)$$

$$+ 2\lambda_h(\lambda_h + \lambda_{hu} + \lambda_c + \mu_2) \ ,$$

$$D_{10} = (\lambda_h + \lambda_{hu} + \lambda_c + \mu_1)[(2\lambda_h + 2\lambda_{hu} + \lambda_{c1})(\lambda_h + \lambda_{hu} + \lambda_c + \mu_2)$$

$$- 2\lambda_{hu}\mu_2] - 2\lambda_h\mu_1(\lambda_h + \lambda_{hu} + \lambda_c + \mu_2) \ .$$

$\mathrm{MTTF_R}$ is the mean time to failure of the system with repair.

Setting $\mu_1 = \mu_2 = 0$ in Eq. (6.181) yields the mean time to failure without repair:

$$\mathrm{MTTF} = [3(\lambda_h + \lambda_{hu}) + \lambda_c][(2\lambda_h + 2\lambda_{hu} + \lambda_{c1})(\lambda_h + \lambda_{hu} + \lambda_c)]^{-1} \ . \tag{6.182}$$

EXAMPLE 6.6

A parallel system is composed of two independent, identical and active units. A unit fails either due to a noncritical human error or due to a hardware failure. The occurrence of critical human error at any system operating state causes total system failure. The unit noncritical human error and hardware failure rates are $\lambda_{hu} = 0.0006$ error/hr and $\lambda_h = 0.002$ failure/hr, respectively. When both units are operating normally, the constant critical human error rate is 0.0003 error/hr. Similarly, when only one unit is operating normally the constant critical human error rate is 0.0001 error/hr. Compute the value of the system mean time to failure with the aid of Eq. (6.182).

In this example the data for the following parameters are specified: $\lambda_{hu} = 0.0006$ error/hr, $\lambda_h = 0.002$ failure/hr, $\lambda_c = 0.0001$ error/hr, and $\lambda_{cl} = 0.0003$ error/hr.

With the aid of Eq. (6.182) and the above data we get

$$\text{MTTF} = \frac{3(0.002 + 0.0006) + 0.0001}{[2(0.002) + 2(0.0006) + 0.0003](0.002 + 0.0006 + 0.0001)}$$

$$= 531.99 \text{ hr} .$$

Thus the system mean time to failure is 531.99 hr.

RELIABILITY ANALYSIS OF ON-SURFACE TRANSIT SYSTEMS WITH HUMAN ERRORS

This section presents four Markov models to evaluate the reliability and availability of transit systems with human errors.

Model I

This model represents a repairable on-surface transit system in which the failures due to human error and the failures due to hardware failures are separated [7]. More specifically, in the field, when the vehicle is operating normally, it can fail completely either due to human error or due to hardware failure. In either of these two cases, a repairman is sent to the failed vehicle's site. If the repairman is able to repair the failed vehicle completely then it is put back into its normal operation; otherwise it is towed to the repair workshop. The fully repaired vehicle is put back into its normal operation. The system transition diagram is shown in Fig. 6.14. The next model is the special case of this model when there is no repair.

The following assumptions are associated with this model.

1. Failure rates are constant.
2. Failures are statistically independent.
3. Repair rates are constant and the repaired vehicle is as good as new.
4. The vehicle can fail completely either due to human errors or due to hardware failures.
5. The towing rates are constant.

The system states are

0   vehicle operating,
1   vehicle failed in the field due to human errors,
2   vehicle failed in the field due to hardware failure,
3   vehicle in the repair shop.

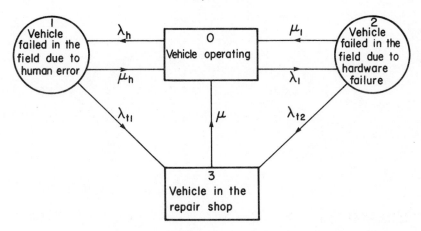

Figure 6.14. System transition diagram for model I.

The notations associated with this model are: $P_i(t)$ is the probability that the vehicle is in state i at time t, $i = 0, 1, 2, 3$. λ_h, λ_1 are the vehicle failure rates from state 0 to 1 and 0 to 2, respectively. λ_{t1}, λ_{t2} are the towing rates from state 1 to 3 and 2 to 3, respectively. μ_h, μ_1, μ are the vehicle repair rates from states 1, 2, 3 to 0, respectively. AV_{ss} is the steady-state availability of the vehicle.

The system of differential equations associated with Fig. 6.14 is

$$\frac{dP_0(t)}{dt} + (\lambda_h + \lambda_1)P_0(t) = P_1(t)\mu_h + P_2(t)\mu_1 + P_3(t)\mu \ , \qquad (6.183)$$

$$\frac{dP_1(t)}{dt} + (\lambda_{t1} + \mu_h)P_1(t) = P_0(t)\lambda_h \ , \qquad (6.184)$$

$$\frac{dP_2(t)}{dt} + (\lambda_{t2} + \mu_1)P_2(t) = P_0(t)\lambda_1 \ , \qquad (6.185)$$

$$\frac{dP_3(t)}{dt} + \mu P_3(t) = P_1(t)\lambda_{t1} + P_2(t)\lambda_{t2} \ . \qquad (6.186)$$

At $t = 0$, $P_0(0) = 1$, and other initial condition probabilities are equal to zero.

By setting the derivatives equal to zero in Eqs. (6.183)–(6.186) and using the relationship $\sum_{i=0}^{3} P_i = 1$, the resulting system steady-state probability equations are

$$P_0 = N_1/D_1 \tag{6.187}$$

and

$$P_i = k_i P_0 \qquad \text{for } i = 1,2,3 \ , \tag{6.188}$$

where

$$N_1 = \mu(\lambda_{t1} + \mu_h)(\lambda_{t2} + \mu_1) \ ,$$

$$D_1 = (\lambda_{t1} + \mu_h)[\mu(\lambda_{t2} + \mu_1) + \lambda_1(\mu + \lambda_{t2})] + \lambda_h(\lambda_{t2} + \mu_1)(\lambda_{t1} + \mu) \ ,$$

$$k_1 = \lambda_h/(\lambda_{t1} + \mu_h) \ ,$$

$$k_2 = \lambda_1/(\lambda_{t2} + \mu_1) \ ,$$

$$k_3 = \frac{\lambda_h \lambda_{t1}(\lambda_{t2} + \mu_1) + \lambda_1 \lambda_{t2}(\lambda_{t1} + \mu_h)}{\mu(\lambda_{t1} + \mu_h)(\lambda_{t2} + \mu_1)} \ .$$

The steady-state availability of the system is

$$AV_{ss} = P_0 \ . \tag{6.189}$$

The steady-state availability plots of Eq. (6.189) are shown in Fig. 6.15. The plots in Fig. 6.15 indicate that for specified values of λ_1, λ_{t2}, μ_h, μ_1 and μ, the AV_{ss} decreases for increasing values of λ_{t1}. In addition, AV_{ss} decreases as the value of λ_h increases.

Model II

This is a special case of model I shown in Fig. 6.14 when $\mu_h = \mu_1 = \mu = 0$. The system of differential equations associated with this model is

$$\frac{dP_0(t)}{dt} + (\lambda_h + \lambda_1)P_0(t) = 0 \ , \tag{6.190}$$

$$\frac{dP_1(t)}{dt} + \lambda_{t1}P_1(t) = P_0(t)\lambda_h \ , \tag{6.191}$$

$$\frac{dP_2(t)}{dt} + \lambda_{t2}P_2(t) = P_0(t)\lambda_1 \ , \tag{6.192}$$

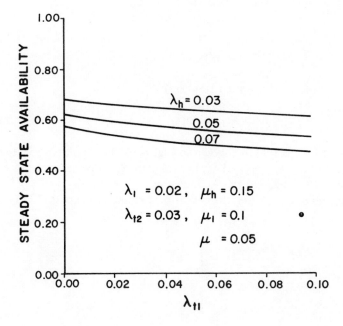

Figure 6.15. Vehicle steady-state availability plots for model I.

$$\frac{dP_3(t)}{dt} = P_1(t)\lambda_{t1} + P_2(t)\lambda_{t2} \ . \tag{6.193}$$

At time $t = 0$, $P_0(0) = 1$, $P_1(0) = P_2(0) = P_3(0) = 0$.

With the aid of Laplace transforms we find from Eqs. (6.190)–(6.194) the resulting probability equations:

$$P_0(t) = e^{-bt} \ , \tag{6.194}$$

$$P_1(t) = b_1(e^{-bt} - e^{-\lambda_{t1}t}) \ , \tag{6.195}$$

$$P_2(t) = b_2(e^{-bt} - e^{-\lambda_{t2}t}) \ , \tag{6.196}$$

$$P_3(t) = 1 + b_1e^{-\lambda_{t1}t} + b_2e^{-\lambda_{t2}t} + b_3e^{-bt} \ , \tag{6.197}$$

where

$$b = \lambda_h + \lambda_l \ ,$$

$$b_1 = \lambda_h/(\lambda_{t1} - b) \ ,$$

$$b_2 = \lambda_1/(\lambda_{t2} - b) \; ,$$

$$b_3 = -(b_1\lambda_{t1} + b_2\lambda_{t2})/b \; .$$

The reliability of the system is given by

$$R(t) = P_0(t) \; . \tag{6.198}$$

The plots of Eq. (6.198) are shown in Fig. 6.16. These plots show that for given values of λ_1, λ_{t1}, λ_{t2} and λ_h, the value of $R(t)$ decreases as t increases. In addition, $R(t)$ decreases as λ_h increases.

With the aid of Eq. (6.198) the mean time to system failure (MTTF) is given by

$$\text{MTTF} = \int_0^\infty R(t)\, dt$$

$$= \frac{1}{b} \; . \tag{6.199}$$

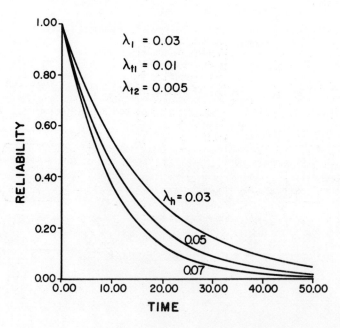

Figure 6.16. Vehicle reliability plots for model II.

Model III

This Markov model represents a transit system which can either fail safely or fail with accident due to hardware failures or human errors [8]. The failed system is taken to the repair shop. The repaired vehicle is put back into normal operation. The state-space diagram of the model is shown in Fig. 6.17. The following assumptions are associated with the model.

1. Failure, repair and towing rates are constant.
2. Human error rates are constant.
3. Failures are statistically independent.
4. A repaired vehicle is as good as new.

The following symbols are associated with this model (the numerals in the boxes of Fig. 6.17 represent the state number). λ_{h1} is the hardware failure rate of the vehicle failing safely. λ_{h2} is the hardware failure rate of the vehicle, causing an accident. λ_{hu1} is the vehicle safe-failure human-error rate. λ_{hu2} is the human error rate of the vehicle, causing an accident. λ_i is the towing rate from state i to state 5, for $i = 1,2,3,4$. μ is the transition rate (repair rate) from state 5 to state 0. s is the Laplace transform variable. $P_i(t)$ is the probability that the vehicle is in state i at time t, for $i = 0,1,2,3,4,5$.

The system of differential equations associated with Fig. 6.17 is

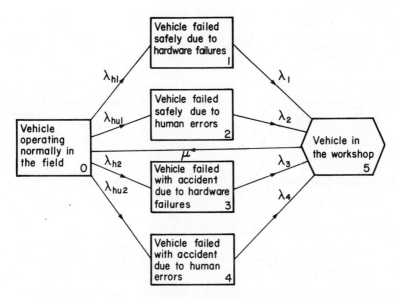

Figure 6.17. State-space diagram for transit system model III.

$$\frac{dP_0(t)}{dt} + (\lambda_{h1} + \lambda_{h2} + \lambda_{hu1} + \lambda_{hu2})P_0(t) = P_5(t)\mu \ , \qquad (6.200)$$

$$\frac{dP_1(t)}{dt} + \lambda_1 P_1(t) = P_0(t)\lambda_{h1} \ , \qquad (6.201)$$

$$\frac{dP_2(t)}{dt} + \lambda_2 P_2(t) = P_0(t)\lambda_{hu1} \ , \qquad (6.202)$$

$$\frac{dP_3(t)}{dt} + \lambda_3 P_3(t) = P_0(t)\lambda_{h2} \ , \qquad (6.203)$$

$$\frac{dP_4(t)}{dt} + \lambda_4 P_4(t) = P_0(t)\lambda_{hu2} \ , \qquad (6.204)$$

$$\frac{dP_5(t)}{dt} + \mu P_5(t) = P_1(t)\lambda_1 + P_2(t)\lambda_2 + P_3(t)\lambda_3 + P_4(t)\lambda_4 \ . \qquad (6.205)$$

At time $t = 0$, $P_0(0) = 1$, and all other initial condition probabilities are equal to zero.

Setting $\mu = 0$ in Eqs. (6.200)–(6.205), taking Laplace transforms and solving for $P_0(s)$ yields

$$R(s) = P_0(s) = \frac{1}{s + A_1} \ , \qquad (6.206)$$

where $A_1 = \lambda_{h1} + \lambda_{h2} + \lambda_{hu1} + \lambda_{hu2}$ and $R(s)$ is the Laplace transform of the transit system reliability function. Thus the transit system reliability is given by

$$R(t) = e^{-A_1 t} \ . \qquad (6.207)$$

Mean time to failure (MTTF) of the transit system is given by

$$\text{MTTF} = \lim_{s \to 0} R(s) = \frac{1}{A_1} \ . \qquad (6.208)$$

Setting the derivatives of Eqs. (6.200)–(6.205) equal to zero and utilizing the relationship $\sum_{i=0}^{5} P_i = 1$ leads to the following steady-state probability solutions:

$$P_0 = \frac{1}{1 + A_2} \tag{6.209}$$

where

$$A_2 = \frac{\lambda_{h1}}{\lambda_1} + \frac{\lambda_{hu1}}{\lambda_2} + \frac{\lambda_{h2}}{\lambda_3} + \frac{\lambda_{hu2}}{\lambda_4} + \frac{A_1}{\mu}$$

(A_1 is already specified),

$$P_1 = \frac{\lambda_{h1}}{\lambda_1} P_0 \ , \tag{6.210}$$

$$P_2 = \frac{\lambda_{hu1}}{\lambda_2} P_0 \ , \tag{6.211}$$

$$P_3 = \frac{\lambda_{h2}}{\lambda_3} P_0 \ , \tag{6.212}$$

$$P_4 = \frac{\lambda_{hu2}}{\lambda_4} P_0 \ , \tag{6.213}$$

$$P_5 = \frac{A_1}{\mu} P_0 \ . \tag{6.214}$$

The steady-state availability of the transit system is given by

$$AV_{ss} = P_0 \ . \tag{6.215}$$

The plots of Eq. (6.215) are shown in Fig. 6.18. These plots clearly show the effect of human error on transit system steady-state availability. From these plots, it is observed that steady-state availability of the transit system decreases with the increasing values of λ_{hu1} and λ_{hu2}.

The steady-state probability of the vehicle failing safely is given by

$$P_s = P_1 + P_2 \ . \tag{6.216}$$

Similarly, the steady-state probability of the vehicle causing an accident is given by

$$P_A = P_3 + P_4 \ . \tag{6.217}$$

The steady state-probability of vehicle failure due to hardware failures is

$$P_h = P_1 + P_3 \ . \tag{6.218}$$

Finally, the steady-state probability of vehicle failure due to human errors is

$$P_{hu} = P_2 + P_4 \ . \tag{6.219}$$

EXAMPLE 6.7

In Fig. 6.17, values for transition rates are as follows: $\lambda_{h1} = 0.004$ failure/hr, $\lambda_{hu1} = 0.0004$ error/hr, $\lambda_{h2} = 0.0005$ failure/hr, and $\lambda_{hu2} = 0.0001$ error/hr.

Calculate the value of the transit system mean time to failure with the aid of Eq. (6.208).

From Eq. (6.208) we get

$$\text{MTTF} = \frac{1}{A_1} = \frac{1}{\lambda_{h1} + \lambda_{h2} + \lambda_{hu1} + \lambda_{hu2}}$$

$$= \frac{1}{(0.004) + (0.0005) + (0.0004) + (0.0001)}$$

$$= 200 \text{ hr} \ .$$

Model IV

This model is also concerned with an on-surface transit system subject to hardware failures and human errors [8]. However, in this case the operating vehicle (i.e., transit system) can degrade/fail either due to hardware failures or to human errors. Completely or seriously partially failed vehicles are taken to the repair shop. The degraded vehicle may fail either due to hardware failures or to human errors. The repaired vehicle is put back into normal operation. The state-space diagram of the system is shown in Fig. 6.19.

The following assumptions are associated with this model.

1. Failure, repair, towing and other transition rates are constant.
2. Human error rates are constant.
3. Failures are statistically independent.
4. The repaired vehicle is as good as new.

The following symbols are associated with this model (numerals in Fig. 6.19 denote corresponding states). λ_{h1} is the vehicle degradation rate due to hardware failures. λ_{h2} is the hardware failure rate of the vehicle from state 1 to state 3. λ_{h3} is the hardware failure rate of the vehicle from state 2 to state 3. λ_h is the hardware failure rate of the vehicle from state 0 to state 3.

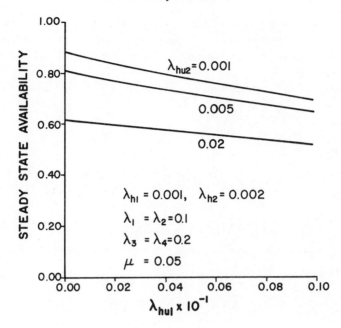

Figure 6.18. Steady-state availability plots for Eq. (6.215).

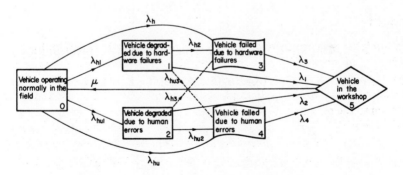

Figure 6.19. State-space diagram for transit system model IV.

λ_{hu1} is the vehicle degradation rate due to human errors. λ_{hu2} is the failure rate of the vehicle due to human errors from state 2 to state 4. λ_{hu3} is the failure rate of the vehicle due to human errors from state 1 to state 4. λ_{hu} is the failure rate of the vehicle due to human errors from state 0 to state 4. λ_i is the vehicle transition rate from state i to state 5, for $i = 1,2,3,4$. (λ_3 and

λ_4 are towing rates from states 3 and 4, respectively, to state 5.) μ is the transition rate (repair rate) from state 5 to state 0. s is the Laplace transform variable. $P_i(t)$ is the probability that the system is in state i at time t, for $i = 0,1,2,3,4,5$.

The system of differential equations associated with Fig. 6.19 is

$$\frac{dP_0(t)}{dt} + (\lambda_h + \lambda_{hu} + \lambda_{h1} + \lambda_{hu1})P_0(t) = P_5(t)\mu \ , \tag{6.220}$$

$$\frac{dP_1(t)}{dt} + (\lambda_{h2} + \lambda_{hu3} + \lambda_1)P_1(t) = P_0(t)\lambda_{h1} \ , \tag{6.221}$$

$$\frac{dP_2(t)}{dt} + (\lambda_{hu2} + \lambda_{h3} + \lambda_2)P_2(t) = P_0(t)\lambda_{hu1} \ , \tag{6.222}$$

$$\frac{dP_3(t)}{dt} + \lambda_3 P_3(t) = P_0(t)\lambda_h + P_1(t)\lambda_{h2} + P_2(t)\lambda_{h3} \ , \tag{6.223}$$

$$\frac{dP_4(t)}{dt} + \lambda_4 P_4(t) = P_0(t)\lambda_{hu} + P_2(t)\lambda_{hu2} + P_1(t)\lambda_{hu3} \ , \tag{6.224}$$

$$\frac{dP_5(t)}{dt} + \mu P_5(t) = P_1(t)\lambda_1 + P_2(t)\lambda_2 + P_3(t)\lambda_3 + P_4(t)\lambda_4 \ . \tag{6.225}$$

At time $t = 0$, $P_0(0) = 1$, and all other initial condition probabilities are equal to zero.

By setting $\mu = 0$ in Eqs. (6.220)–(6.225), taking Laplace transforms and solving for $P_0(s)$, $P_1(s)$, and $P_2(s)$ yields

$$R(s) = P_0(s) + P_1(s) + P_2(s) \ , \tag{6.226}$$

where $R(s)$ is the Laplace transform of the transit system reliability function,

$$P_0(s) = \frac{1}{s + B_1} \ , \tag{6.227}$$

$$B_1 = \lambda_h + \lambda_{hu} + \lambda_{h1} + \lambda_{hu1} \ ,$$

$$P_1(s) = \frac{\lambda_{h1}}{(s + B_1)(s + B_2)} \ , \tag{6.228}$$

$$B_2 = \lambda_{h2} + \lambda_{hu3} + \lambda_1 \ ,$$

$$P_2(s) = \frac{\lambda_{hu1}}{(s + B_1)(s + B_3)} \ , \tag{6.229}$$

$$B_3 = \lambda_{hu2} + \lambda_{h3} + \lambda_2 \ .$$

With the aid of Eq. (6.226) the reliability of the transit system is given by

$$R(t) = e^{-B_1 t} + \frac{\lambda_{h1}}{B_2 - B_1} \left(e^{-B_1 t} - e^{-B_2 t} \right) + \frac{\lambda_{hu1}}{B_3 - B_1} \left(e^{-B_1 t} - e^{-B_3 t} \right) \ . \tag{6.230}$$

Mean time to failure (MTTF) of the transit system is given by

$$\text{MTTF} = \lim_{s \to 0} R(s) = \frac{1}{B_1} \left(1 + \frac{\lambda_{h1}}{B_2} + \frac{\lambda_{hu1}}{B_3} \right) \ . \tag{6.231}$$

The plots of Eq. (6.231) are shown in Fig. 6.20. These plots show the effect of λ_{hu1} and λ_{h1} on the MTTF of the transit system. Furthermore, it is observed that the MTTF decreases with increasing values of λ_{hu1} and λ_{h1}.

Setting the derivatives of Eqs. (6.220)–(6.225) equal to zero and utilizing the relationship $\sum_{i=0}^{5} P_i = 1$ leads to the following steady-state probability solutions:

$$P_0 = \frac{1}{1 + B_7} \ , \tag{6.232}$$

where B_1, B_2, B_3 have already been defined,

$$B_4 = \frac{1}{\lambda_3} \left(\lambda_h + \frac{\lambda_{h1} \lambda_{h2}}{B_2} + \frac{\lambda_{hu1} \lambda_{h3}}{B_3} \right) \ ,$$

$$B_5 = \frac{1}{\lambda_4} \left(\lambda_{hu} + \frac{\lambda_{hu1} \lambda_{hu2}}{B_3} + \frac{\lambda_{h1} \lambda_{hu3}}{B_2} \right) \ ,$$

$$B_6 = \frac{1}{\mu} \left(\frac{\lambda_1 \lambda_{h1}}{B_2} + \frac{\lambda_2 \lambda_{hu1}}{B_3} + \lambda_3 B_4 + \lambda_4 B_5 \right) \ ,$$

$$B_7 = \frac{\lambda_{h1}}{B_2} + \frac{\lambda_{hu1}}{B_3} + B_4 + B_5 + B_6 \ ,$$

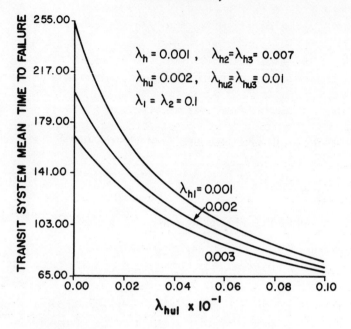

Figure 6.20. Mean time to failure plots for model IV.

$$P_1 = \frac{\lambda_{h1}}{B_2} P_0 \ , \tag{6.233}$$

$$P_2 = \frac{\lambda_{hu1}}{B_3} P_0 \ , \tag{6.234}$$

$$P_3 = B_4 P_0 \ , \tag{6.235}$$

$$P_4 = B_5 P_0 \ , \tag{6.236}$$

$$P_5 = B_6 P_0 \ . \tag{6.237}$$

The steady-state availability of the transit system is given by

$$AV_{ss} = P_0 + P_1 + P_2 \ . \tag{6.238}$$

RELIABILITY EVALUATION OF HUMAN OPERATORS UNDER STRESS

This section presents two models concerned with reliability evaluation of human operators performing time-continuous tasks under fluctuating stress.

Model I

This model is concerned with a four-state Markov model [9]. The state-space diagram of the model is shown in Fig. 6.21. In this model, the human operator performing the time-continuous task fluctuates between normal work and stress states. The task is associated with a system. The system fails due to a human error from either the normal work state or the stress state (when the operator is performing under normal conditions or under stress).

The following assumptions are associated with this model:

1. Errors are statistically independent.
2. Human error rates are constant.
3. The human operator is performing a time-continuous task.
4. The rate of changing human operator condition from the normal state to the stress state and vice versa is constant.
5. Numerals in Fig. 6.21 denote corresponding states.

The following symbols are associated with this model: λ_1 is the constant human error rate from state 0. λ_2 is the constant human error rate from state 2. α is the transition rate from the normal state to the stress state. β is the transition rate from the stress state to the normal state. s is the Laplace transform variable. $P_i(t)$ is the probability of being in state i at time t, for $i = 0,1,2,3$.

The system of differential equations associated with Fig. 6.21 is

$$\frac{dP_0(t)}{dt} + (\lambda_1 + \alpha)P_0(t) = P_2(t)\beta , \qquad (6.239)$$

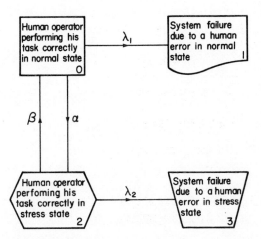

Figure 6.21. State-space diagram for the human operator under alternating stress.

$$\frac{dP_1(t)}{dt} = P_0(t)\lambda_1 \ , \tag{6.240}$$

$$\frac{dP_2(t)}{dt} + (\lambda_2 + \beta)P_2(t) = P_0(t)\alpha \ , \tag{6.241}$$

$$\frac{dP_3(t)}{dt} = P_2(t)\lambda_2 \ . \tag{6.242}$$

At time $t = 0$, $P_0(0) = 1$ and $P_1(0) = P_2(0) = P_3(0) = 0$.

With the aid of the Laplace transformation technique, from equations (6.239)–(6.242) the resulting state probability equations can be derived:

$$P_0(t) = (x_2 - x_1)^{-1}[(x_2 + \lambda_2 + \beta)e^{x_2 t} - (x_1 + \lambda_2 + \beta)e^{x_1 t}] \ , \tag{6.243}$$

where

$$x_1 = \frac{-b_1 + \sqrt{b_1^2 - 4b_2}}{2} \ ,$$

$$x_2 = \frac{-b_1 - \sqrt{b_1^2 - 4b_2}}{2} \ ,$$

$$b_1 = \lambda_1 + \lambda_2 + \alpha + \beta \ ,$$

$$b_2 = \lambda_1(\lambda_2 + \beta) + \alpha\lambda_2 \ ;$$

$$P_1(t) = b_4 + b_5 e^{x_2 t} - b_6 e^{x_1 t} \ , \tag{6.244}$$

where

$$b_3 = \frac{1}{x_2 - x_1} \ ,$$

$$b_4 = \lambda_1(\lambda_2 + \beta)/x_1 x_2 \ ,$$

$$b_5 = b_3(\lambda_1 + b_4 x_1) \ ,$$

$$b_6 = b_3(\lambda_1 + b_4 x_2) \ ;$$

$$P_2(t) = \alpha b_3(e^{x_2 t} - e^{x_1 t}) \ ; \tag{6.245}$$

$$P_3(t) = b_7[(1 + b_3)(x_1 e^{x_2 t} - x_2 e^{x_1 t})] \ , \tag{6.246}$$

where

$$b_7 = \lambda_2 \alpha / x_1 x_2 \ .$$

Plots of Eqs. (6.243)–(6.246) are shown in Fig. 6.22 for the specified values of λ_1, λ_2, α, β and t. These plots exhibit the time-dependent behavior of the state probabilities. The human operator reliability is given by

$$R(t) = P_0(t) + P_2(t) \ . \tag{6.247}$$

The mean time to human error is given by

$$\text{MTTHE} = \int_0^\infty R(t) \, \mathrm{d}t$$

$$= \int_0^\infty [P_0(t) + P_2(t)] \, \mathrm{d}t$$

$$= (\lambda_2 + \alpha + \beta)/b_2 \ . \tag{6.248}$$

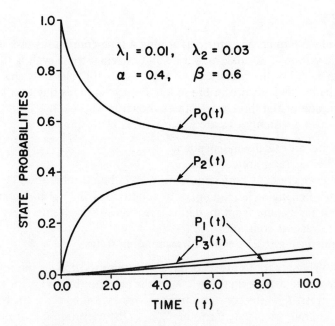

Figure 6.22. State probability plots for model I.

EXAMPLE 6.8

A person has to perform a time-continuous task under normal and stress conditions. In the normal condition the constant human error rate is 0.01 error/hr. Similarly, under the stress condition the human-error rate is 0.04 error/hr. The value of the transition rate from normal condition to stress condition is 0.05 hr^{-1}, and conversely, 0.02 hr^{-1}. Calculate the value of the mean time to human error.

By substituting the given data into Eq. (6.248) we get

$$\text{MTTHE} = \frac{\lambda_2 + \alpha + \beta}{b_2}$$

$$= \frac{\lambda_2 + \alpha + \beta}{\lambda_1(\lambda_2 + \beta) + \alpha\lambda_2}$$

$$= \frac{0.04 + 0.05 + 0.02}{0.01(0.04 + 0.02) + (0.05)(0.04)}$$

$$= \frac{0.11}{0.0006 + 0.002} = 42.31 \text{ hr} .$$

Thus the mean time to human error is 42.31 hr.

Model II

In this model a human operator performing a time-continuous task associated with a system at normal, moderate and extreme stress levels is studied. The above model is the special case of this model [10]. The state-space diagram of the model is shown in Fig. 6.23. The system fails due to a human error from any of the three work states shown in Fig. 6.23.

The following assumptions are associated with this model:

1. Errors are statistically independent.
2. Human error rates are constant.
3. A human operator is performing a time-continuous operation task.
4. The rate of changing human operator conditions from normal to moderate stress levels, and from moderate stress levels to extreme stress levels and vice versa is constant.
5. Numerals in Fig. 6.23 denote corresponding states.

The following notations are associated with this model. λ_1 is the constant human error rate in a normal state. λ_2 is the constant human error rate at moderate stress. λ_3 is the constant human error rate at extreme stress. α_1 is the transition rate from state 0 to state 2. α_2 is the transition rate from state 2 to state 4. α_3 is the transition rate from state 0 to state 4. β_1 is the transi-

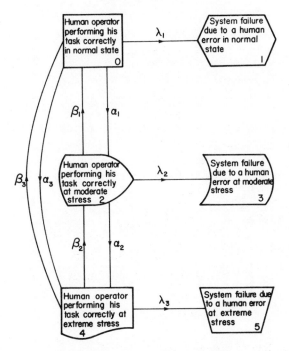

Figure 6.23. State-space diagram for model II.

tion rate from state 2 to state 0. β_2 is the transition rate from state 4 to state 2. β_3 is the transition rate from state 4 to state 0. s is the Laplace transform variable. $P_i(t)$ is the probability of being in state i at time t, for $i = 0,1,2,3,4,5$.

The system of differential equations associated with Fig. 6.23 is

$$\frac{dP_0(t)}{dt} + (\lambda_1 + \alpha_1 + \alpha_3)P_0(t) = P_2(t)\beta_1 + P_4(t)\beta_3 , \qquad (6.249)$$

$$\frac{dP_1(t)}{dt} = P_0(t)\lambda_1 , \qquad (6.250)$$

$$\frac{dP_2(t)}{dt} + (\lambda_2 + \alpha_2 + \beta_1)P_2(t) = P_0(t)\alpha_1 + P_4(t)\beta_2 , \qquad (6.251)$$

$$\frac{dP_3(t)}{dt} = P_2(t)\lambda_2 , \qquad (6.252)$$

$$\frac{dP_4(t)}{dt} + (\lambda_3 + \beta_2 + \beta_3)P_4(t) = P_2(t)\alpha_2 + P_0(t)\alpha_3 \ , \tag{6.253}$$

$$\frac{dP_5(t)}{dt} = P_4(t)\lambda_3 \ . \tag{6.254}$$

At time $t = 0$, $P_0(0) = 1$, and all other initial condition probabilities are equal to zero.

With the aid of Laplace transforms, from equations (6.249)–(6.254) the Laplace transforms of state probabilities can be found as follows:

$$P_0(s) = \frac{(s + k_2)(s^2 + k_5 s + k_6)}{s^4 + k_{12}s^3 + k_{13}s^2 + k_{14}s + k_{15}} \ , \tag{6.255}$$

where

$$k_1 = \lambda_1 + \alpha_1 + \alpha_3 \ ,$$

$$k_2 = \lambda_2 + \alpha_2 + \beta_1 \ ,$$

$$k_3 = \lambda_3 + \beta_2 + \beta_3 \ ,$$

$$k_4 = \alpha_3 k_2 + \alpha_1 \alpha_2 \ ,$$

$$k_5 = k_2 + k_3 \ ,$$

$$k_6 = k_2 k_3 + \alpha_2 \beta_2 \ ,$$

$$k_7 = \alpha_1 \beta_1 + \alpha_3 \beta_3 \ ,$$

$$k_8 = \alpha_1 \beta_1 k_5 + \beta_1 \beta_2 \alpha_3 + \beta_3 k_4 + \alpha_3 \beta_3 k_2 \ ,$$

$$k_9 = \alpha_1 \beta_1 k_6 + \beta_1 \beta_2 k_4 + k_2 k_4 \beta_3 \ ,$$

$$k_{10} = k_1 + k_2 \ ,$$

$$k_{11} = k_1 k_2 \ ,$$

$$k_{12} = k_5 + k_{10} \ ,$$

$$k_{13} = k_6 + k_{11} + k_5 k_{10} - k_7 \ ,$$

$$k_{14} = k_5 k_{11} + k_6 k_{10} - k_8 \ ,$$

$$k_{15} = k_6 k_{11} - k_9 \ ;$$

$$P_1(s) = \frac{\lambda_1}{s} P_0(s) \ ; \tag{6.256}$$

$$P_2(s) = \frac{\alpha_1(s^2 + k_5 s + k_6) + \beta_2(\alpha_3 s + k_4)}{(s + k_2)(s^2 + k_5 s + k_6)} P_0(s) \ ; \tag{6.257}$$

$$P_3(s) = \frac{\lambda_2}{s} P_2(s) \ ; \tag{6.258}$$

$$P_4(s) = \frac{\alpha_3 s + k_4}{s^2 + k_5 s + k_6} P_0(s) \ ; \tag{6.259}$$

$$P_5(s) = \frac{\lambda_3}{s} P_4(s) \ . \tag{6.260}$$

The Laplace transform of the operator reliability is given by

$$R(s) = P_0(s) + P_2(s) + P_4(s) \ . \tag{6.261}$$

The mean time to human error is given by

$$\text{MTTHE} = \lim_{s \to 0} R(s) = \frac{k_2 k_6}{k_{15}} \left[1 + \frac{\alpha_1 k_6 + \beta_2 k_4}{k_2 k_6} + \frac{k_4}{k_6} \right] \ . \tag{6.262}$$

The plots of Eq. (6.262) are shown in Fig. 6.24. These plots show the effect of λ_1 and λ_3 on the mean time to human error for specific values of the parameters. It is evident from the plots that the MTTHE decreases for the increasing values of λ_1 and λ_3.

AVAILABILITY ANALYSIS OF A SYSTEM WITH HUMAN ERRORS

This section presents a model representing a system that may fail due to human errors or general (hardware and other) failures [11]. Due to any one of these failures the operational system may be degraded or stop operating (fail). The system may fail from its degraded states due to general failures or human errors.

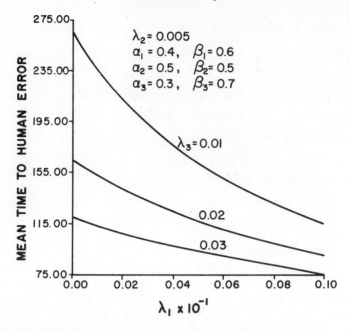

Figure 6.24. Mean time to human error plots for model II.

The system transition diagram is shown in Fig. 6.25. The method of supplementary variables is used to develop equations for the model. This technique is described in detail in Ref. [1].

The following assumptions are associated with this model.

1. Human and other failures are s-independent.
2. Human and other failure rates are constant.
3. The repaired system is as good as new.
4. Failed system repair times are arbitrarily distributed.
5. Degraded system repair rates are constant.
6. The system is repaired from its degraded and failed modes to a good or normal operating state.
7. The degraded system under repair may fail due to human and general (hardware) failures.
8. System states are good (g), degraded due to human errors (dh), degraded due to general (hardware and other) failures (dg), failed due to human errors (fh), and failed due to general (hardware and other) failures (fg).

The following symbols are associated with this model. $P_i(t)$ is the probability that the system is in an unfailed state i at time t for $i = $ g, dg, dh.

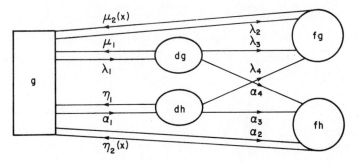

Figure 6.25. System transition diagram.

$p_j(x, t)$ is the probability density (with respect to repair time) that the failed system is in state j and has an elapsed repair time of x for $j =$ fg, fh. $\mu_2(x)$, $q_{fg}(x)$ is the repair hazard rate and pdf of repair time when the system is in state fg and has an elapsed repair time of x. $\eta_2(x)q_{fh}(x)$ is the repair rate (a hazard rate) and pdf of repair time when the system is in state fh and has an elapsed repair time of x. α_i is the constant human error rate i for $i = 1$ (g to dh), $i = 2$ (g to fh), $i = 3$ (dh to fh), $i = 4$ (dg to fh). λ_i is the constant general (hardware, software and other) failure rate i for $i = 1$ (g to dg); $i = 2$ (g to fg), $i = 3$ (dg to fg), $i = 4$ (dh to fg). μ_1 is the constant system repair rate from state dg to state g. η_1 is the constant system repair rate from state dh to state g. s is the Laplace transform variable.

The equations for the model are

$$\frac{dP_g(t)}{dt} + P_g(t)(\lambda_1 + \lambda_2 + \alpha_1 + \alpha_2) = P_{dh}(t)\eta_1 + P_{dg}(t)\mu_1$$

$$+ \int_0^\infty p_{fg}(x, t)\mu_2(x)\, dx$$

$$+ \int_0^\infty p_{fh}(x, t)\eta_2(x)\, dx \;, \quad (6.263)$$

$$\frac{dP_{dg}(t)}{dt} + P_{dg}(t)(\lambda_3 + \alpha_4 + \mu_1) = P_g(t)\lambda_1 \;, \quad (6.264)$$

$$\frac{dP_{dh}(t)}{dt} + P_{dh}(t)(\eta_1 + \alpha_3 + \lambda_4) = P_g(t)\alpha_1 \;, \quad (6.265)$$

$$\frac{\partial p_{fg}(x, t)}{\partial t} + \frac{\partial p_{fg}(x, t)}{\partial x} + p_{fg}(x, t)\mu_2(x) = 0 \;, \quad (6.266)$$

$$\frac{\partial p_{\text{fh}}(x,t)}{\partial t} + \frac{\partial p_{\text{fh}}(x,t)}{\partial x} + p_{\text{fh}}(x,t)\eta_2(x) = 0 \ , \tag{6.267}$$

$$p_{\text{fg}}(0,t) = P_{\text{g}}(t)\lambda_2 + P_{\text{dg}}(t)\lambda_3 + P_{\text{dh}}(t)\lambda_4 \ , \tag{6.268}$$

$$p_{\text{fh}}(0,t) = P_{\text{dg}}(t)\alpha_4 + P_{\text{dh}}(t)\alpha_3 + P_{\text{g}}(t)\alpha_2 \ . \tag{6.269}$$

At $t = 0$, $P_{\text{g}}(0) = 1$, $P_{\text{dg}}(0) = P_{\text{dh}}(0) = p_{\text{fg}}(x,0) = p_{\text{fh}}(x,0) = 0$.
The Laplace transforms of the solutions of (6.263)–(6.267) are

$$P_{\text{g}}(s) = [s + \lambda_1 + \lambda_2 + \alpha_1 + \alpha_2 - (\eta_1\alpha_1/A_1) - (\mu_1\lambda_1/A_2)$$
$$- \{\lambda_2 + (\lambda_3\lambda_1/A_2) + (\lambda_4\alpha_1/A_1)\}G_{\text{fg}}(s)$$
$$- \{(\alpha_4\lambda_1/A_2) + (\alpha_3\alpha_1/A_1) + \alpha_2\}G_{\text{fh}}(s)]^{-1} \ , \tag{6.270}$$

where

$$G_i(s) \equiv \int_0^\infty e^{-sx}q_i\,dx \ , \qquad \text{for } i = \text{fg or fh} \ , \tag{6.271}$$

$$A_1 \equiv s + \lambda_4 + \alpha_3 + \eta_1 \ , \tag{6.272}$$

$$A_2 \equiv s + \lambda_3 + \alpha_4 + \mu_1 \ ; \tag{6.273}$$

$$P_{\text{dg}}(s) = P_{\text{g}}(s)\ \lambda_1/A_2 \ ; \tag{6.274}$$

$$P_{\text{dh}}(s) = P_{\text{g}}(s)\alpha_1/A_1 \ ; \tag{6.275}$$

$$P_{\text{fg}}(s) = [P_{\text{g}}(s)\lambda_2 + P_{\text{dg}}(s)\lambda_3 + P_{\text{dh}}(s)\lambda_4][1 - G_{\text{fg}}(s)]/s \ ; \tag{6.276}$$

$$P_{\text{fh}}(s) = [P_{\text{dg}}(s)\alpha_4 + P_{\text{dh}}(s)\alpha_3 + P_{\text{g}}(s)\alpha_2][1 - G_{\text{fh}}(s)]/s \ . \tag{6.277}$$

To obtain time-domain solutions, Eqs. (6.270)–(6.277) can be transformed for a given completely failed system repair-time distribution.

SUMMARY

This chapter presents various mathematical models for performing reliability availability analysis of systems with human errors.

Four Markov models are presented to determine reliability, mean time to failure and state probabilities of an independent unit parallel system with human error. A number of Markov models concerned with repairable systems with human error are presented. These models are associated with (i) two-unit parallel systems, (ii) two-out-of-three unit systems, and (iii) two-identical-unit standby systems. The steady-state probability equations are developed for all three of these models.

The occurrence of critical human errors is discussed. Three Markov models of redundant systems with critical human errors are presented. The mean time to failure formulas with repair and without repair are developed for two of the three models.

Markov modeling of on-surface transit systems with human errors is presented. Four of the Markov models presented in the chapter deal with on-surface transit systems. Two Markov models for evaluating the human operator reliability are presented. Reliability and mean time to human error formulas are developed. A mathematical model representing a system with partial failures and human errors is analysed with the aid of the supplementary variables technique.

EXERCISES

1. A parallel system is composed of two independent, identical and active units. Unit constant hardware failure and human error rates are 0.007 failure/hr and 0.0001 error/hr, respectively. Each unit may fail due to a hardware failure or a human error. Calculate the system reliability for a 100 hr mission.
2. In the above exercise assume that the system is composed of five units instead of only two units. If everything else remains the same, calculate the system mean time to failure.
3. For Fig. 6.5 prove that the sum of Laplace transforms of state probabilities is $1/s$; that is, $P_0(s) + P_1(s) + P_2(s) + P_3(s) + P_4(s) + P_5(s) = 1/s$. Assume that in Fig. 6.5, $\mu_3 = \mu_4 = \mu_5 = 0$.
4. Obtain an expression for the reliability of a repairable two-out-of-three unit system. The overall system state-space diagram is shown in Fig. 6.7.
5. Explain the difference between the terms "human error" and "critical human error." Give an example of occurrence of critical human error when unit redundancy is involved.
6. Calculate the reliability of the parallel system defined in Example 6.4, for a 50 hr mission.
7. Assume that in Fig. 6.25 the failed system repair times are exponentially distributed. With the aid of Eqs. (6.270)–(6.277) obtain expressions for system steady-state probabilities (i.e., an expression for each of the five system states).

REFERENCES

1. B. S. Dhillon, *Reliability Engineering in Systems Design and Operation*. Van Nostrand Reinhold, New York (1983).
2. E. W. Hagen (Ed.), Human reliability analysis. *Nuclear Safety* 17, 315–326 (1976).

3. B. S. Dhillon and S. N. Rayapati, Reliability analysis of non-maintained parallel systems subject to hardware failure and human error. *Microelectronics and Reliability* **25** (1), 111–122 (1985).
4. B. S. Dhillon and S. N. Rayapati, Analysis of redundant systems with human errors, in *Proceedings of the Annual Reliability and Maintainability Symposium*, IEEE, New York, pp. 315–321 (1985).
5. B. S. Dhillon and R. B. Misra, Reliability evaluation of systems with critical human error. *Microelectronics and Reliability* **24**, 743–759 (1984).
6. B. S. Dhillon and R. B. Misra, Effect of critical human error on system reliability. *Reliability Engineering* **12** (1985).
7. B. S. Dhillon and S. N. Rayapati, Reliability and availability analysis of on surface transit systems. *Microelectronics and Reliability* **24**, 1029–1033 (1984).
8. B. S. Dhillon and S. N. Rayapati, Reliability evaluation of transportation systems with human errors, in *Proceedings of the IASTED International Conference on Applied Simulation and Modeling*, Montreal, 1985, Acta Press, Anaheim, California (1985).
9. B. S. Dhillon, Stochastic models for predicting human reliability. *Microelectronics and Reliability* **22**, 491–496 (1982).
10. B. S. Dhillon and S. N. Rayapati, Reliability evaluation of human operators under stress. *Microelectronics and Reliability* **25**, 729–752 (1985).
11. B. S. Dhillon, System reliability evaluation models with human error. *IEEE Transactions on Reliability* **32**, 47 (1983).

Chapter 7

Human Factors in Maintenance and Maintainability

INTRODUCTION

Human factors play an important role in maintenance and maintainability. Before we go into a detailed discussion on this topic, it is necessary to draw a distinction between maintenance engineering and maintainability engineering. Maintenance engineering is associated with the technical difficulties of keeping equipment in working order, or repairing a failed unit once the system is being used. Maintainability engineering, on the other hand, is associated with the implementing principles that are fundamental to future system repair while the system is in the design and development or fabrication phase [1].

It was not until the 1950s that attention to human factors was focused on design features of systems that impacted on maintenance. An example of such attention is the *Guide to Design of Electronic Equipment for Maintainability*, prepared by J. D. Folley and J. W. Altman in April 1956 [2]. Human factors engineering and maintainability engineering interface in the three main activity areas [3] as shown in Fig. 7.1.

It is a well-known fact that a significantly large proportion of total human errors occur during the maintenance phase [4]. For example, about 25% of the maintenance events described in 213 problem reports from the field were due to human errors [5]. These human errors led to air defense system failure. Furthermore, according to a missile system study reported in Ref. [6], interviews with line and supervisory personnel indicate many more human-initiated malfunctions than reported in the written reports.

This chapter discusses the various aspects of human factors in maintenance and maintainability in subsequent sections.

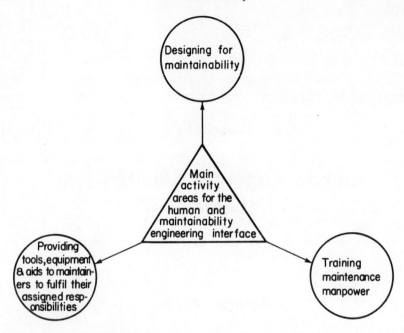

Figure 7.1. Three main activity areas in which human engineering and maintainability engineering interface.

HUMAN FACTORS IN MAINTENANCE

A maintenance man plays a significant role in the reliability of equipment. In the past, catastrophic failures have occurred due to the human element in maintenance. In 1979, the DC-10 accident at Chicago's O'Hare Airport in which 272 persons lost their lives is a prime example. Fault diagnosis takes the majority of the time of the maintenance person. Before the wide application of automatic test equipment, one writer divided a maintenance man's time [7] associated with electronic equipment approximately as shown in Table 7.1.

An example of breakdown percentages of maintenance error causes in missile operations [7] is given with the aid of the histogram in Fig. 7.2. The highest cause of maintenance errors is misreading and missetting of dials and controls.

Various aspects of human factors in maintenance are discussed below.

Training and Experience of Maintenance Personnel

According to one study [8] over a wide variation of tasks such as adjusting, removing and aligning, the average human reliability is 0.987. It means that out of 1000 attempts by maintenance personnel one may expect 13 errors.

Table 7.1. Approximate divisions of a maintenance person's time

Activity	Approximate percentage
1. Fault diagnosis	65–75
2. Remedial actions	15–25
3. Verification	5–15

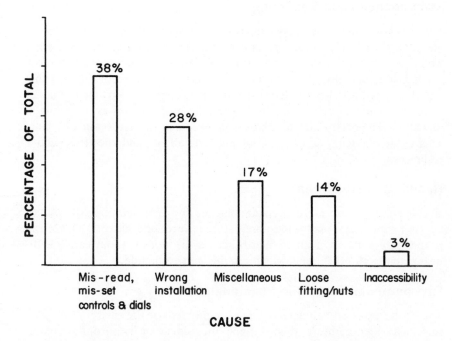

Figure 7.2. Breakdown percentages of maintenance error causes in missile operations.

Obviously, the training and experience of maintenance personnel will also be an important factor in their performance.

Information on characteristics of the trainee manpower for maintenance work is vital to specialists in training and design. According to a study reported in Ref. [8], technicians who ranked high in ratings of their performance had characteristics such as greater aptitude, morale, emotional stability and work experience, and a lesser number of fatigue reports. Furthermore, this study revealed that correlations between characteristics such as morale, experience and responsible handling ability on the one hand, and task performance on the other hand, are positive.

HR–F

Diagnosis Errors

This is an important aspect of maintenance work. Some studies [7] have reported that diagnosis time is not effectively utilized. Many times, parts removed for repair are not actually defective. In one study associated with an aircraft system, it was found that nondefective parts were removed for repair purposes between 40 and 50% of the time. Obviously, this factor will have significant impact on equipment operational reliability.

Maintenance Data Recording

Various shortcomings are associated with descriptions of failures by the operators. One study [7] concerning coded data fields indicated that one error would be present in fewer than half of the forms. The occurrence of error depends almost entirely on the types of characters used. According to Table 7.2, the highest and lowest error rates are associated with alphabetic characters and numbers, respectively. Alphanumerics have an error rate in between these two. The occurrence of errors is independent of factors such as character usage frequency and its position in the data field, and knowledge about equipment.

Handling of Equipment

This is an important factor in maintenance. Improper handling of components or subsystems may lead to premature failures or accidents. Therefore, proper care must be given to factors such as (i) items without handles, (ii) heavy items, and (iii) items which are difficult to handle.

Environmental Factors in Maintenance

The effectiveness of maintenance work is dependent on the environment. Factors such as temperature, dust, fatigue, incomplete or inappropriate maintenance tools, errors in inventory and personal problems may be causes for human errors in maintenance.

HUMAN FACTORS IN MAINTAINABILITY

The human element plays an important role in maintainability engineering tasks and analysis. Major tasks of maintainability engineering are concerned with preparing a program plan and maintainability status reports; taking part in design reviews; performing maintainability demonstrations and analysis; and developing data collection, analysis and corrective measure systems. Most of the components of maintainability analysis [9] are listed below.

1. Establish requirements for maintainability and establish interfaces with reliability, logistics support design, etc.

Table 7.2. Occurrence of errors

Type of codes	Error rate
1. Numbers	lowest
2. Alphabetic characters	highest
3. Alphanumerics	in between

2. Carry out trade-off studies and function analysis.
3. Provide recommendations for spares and perform predictions.
4. Perform allocations and data analysis.
5. Participate in repair policy making and maintenance planning.

Finally, it is emphasized that the above tasks and elements of maintainability analysis demand a considerable amount of attention. Otherwise a greater probability of human-error occurrence will exist.

USEFUL GUIDELINES FOR PROFESSIONALS INVOLVED WITH MAINTENANCE AND MAINTAINABILITY

This section presents maintenance- and maintainability-related guidelines for the electrical designer, maintenance engineer, mechanical designer or maintainability engineer. In addition, expectations from the novice technician and characteristics of a skilled technician are briefly discussed.

Characteristics of a Skilled Maintenance Technician

A maintenance technician has to possess certain characteristics to perform his or her job effectively. The major characteristics [10] are (i) knowledge, (ii) flexibility and (iii) the ability to troubleshoot. If these requirements cannot be met effectively, the design engineer must consider design characteristics such as those below:

1. the throw-away maintenance concept,
2. step-by-step procedures,
3. go/no-go displays,
4. checks at the subsystem level.

Expectations from the Novice Maintenance Technician

Sometimes relatively low-level technicians may be used to perform maintenance. Therefore, this section briefly lists the expectations from novice technicians. Some of these expectations [11] are as follows.

1. Numerous assembly and disassembly errors.
2. Inefficient troubleshooting strategy.
3. Ability to manipulate tool-operated controls, rotary controls, toggle switches, etc.
4. Ability to perform simple arithmetic computations.
5. Unawareness of all potential equipment hazards.
6. Ability to interpret oscilloscope indications.
7. Ability to make use of conventional household tools.
8. Ability to distinguish between basic geometric forms.
9. High-school-level reading and writing abilities.
10. Ability to identify common colors.

Maintenance-Related Guidelines for Electrical and Mechanical Designers

This section presents lists of selected questions for both electrical and mechanical designers [12]. Answers to these questions will ensure better design with respect to maintenance and, in turn fewer maintenance errors.

Electrical designers. These designers should seek answers to questions such as those that follow.

1. Is there any need for special handling of a unit?
2. What kind of adjustments will be required after unit installation in the system?
3. Are the factory and field adjustments minimized to their minimum level?
4. How frequently will the periodic testing be necessary?
5. Are the test points accessible?
6. Are all interconnected circuits contained in the same package?
7. Are parts with high failure rates easily accessible for replacement?
8. Is it possible to replace panel light bulbs without any difficulty?
9. Will the use of a jumper cable be tolerated by a circuit during maintenance work?
10. Is the need for maintenance test equipment minimized?
11. Are components that are subject to early wear-out properly identified?
12. Are the plug pins identified?
13. Is the interaction between circuit parameters and adjustments minimized to an acceptable level?

Mechanical designers. In order to have better equipment maintenance in the field, mechanical designers should seek answers to questions such as those that follow.

1. Is the item design such that the need for special maintenance environments and facilities is kept to a minimum level?
2. Are all items accessible for effective maintenance work?
3. Are all adjustments to be performed in the field accessible?
4. Is necessary protection provided for test and maintenance personnel?
5. Is sufficient access allowed to test points?
6. Were steps taken to avoid complicated-item sequential assembly and disassembly for maintenance work and adjustments?

Guidelines for Maintenance Engineers and Maintainability Engineers

This section lists a number of questions to which both maintenance and maintainability engineers should seek answers during the design phase.

Maintenance engineer. During design the maintenance engineer should seek answers to the following questions [9].

1. Who will maintain the equipment in the field?
2. What skills are to be expected from maintenance personnel?
3. How is the needed maintenance to be integrated into current operations?
4. What requirements of scheduled maintenance are associated with the equipment in question?
5. Are there any special requirements to assure equipment maintenance? If yes, what are they?
6. What plans are needed to fulfill maintainability concerns and time guarantees?
7. What is the operational scenario?
8. Is there a planned operations schedule? If yes, what is it?
9. What is the cost-effective maintenance concept as far as equipment operational requirements?
10. What types of equipment failures have to be verified?
11. What are the concerns of reliability and safety to be considered in maintenance approaches?

Maintainability engineer. In addition to the above questions, the maintainability engineer seeks answers to questions such as these:

1. What are the maintainability performance parameters requiring allocation, prediction and control?
2. Is there any part requiring control to assure interchangeability?
3. What design features are to be specified so that the probability of maintenance error is at a minimum level?

4. What are the probable failure risks, most likely failure causes and their consequences?
5. What design features are necessary to ensure downtime limits?
6. What is the optimum level of repair within the boundaries of downtime?

MAINTENANCE PROCEDURES AND JOB AIDS

This section presents recommendations for preparing maintenance procedures and a procedure for developing job aids. Effective maintenance procedures and job aids will be vital factors in maintenance work with respect to human reliability.

Recommendations for Preparing Maintenance Procedures

In the preparation of maintenance procedures, attention must be paid to recommendations [1] such as those below.

1. Keep the number of decisions to be made by technicians to a minimum level.
2. Make maintenance procedures as short as possible without losing their effectiveness.
3. Follow the step-by-step approaches.
4. Avoid having maintenance personnel work close to delicate parts or dangerous conditions, such as high voltage.
5. Make sure that the procedures clearly state how to start up and shut down the equipment.
6. Develop procedures that provide unambiguous results and are systematic for troubleshooting.
7. Keep the procedures as simple as possible.
8. Reduce the number of alternatives in decision making.

A Method for Developing Job Aids

In maintenance work job aids play an important role. Examples of job aids are schematics, handbooks and manuals. The steps shown in Fig. 7.3 are useful for developing job aids [13].

MANPOWER REQUIREMENT MODELS

This section presents two mathematical models concerned with manpower requirements.

Model I

This model was recommended for estimating total technical manpower requirements for an aircraft configuration [14]. The total manpower, M, is given by

$$M = \sum_{i=1}^{k} m_i \ , \tag{7.1}$$

where m_i is the estimated manpower requirement to support equipment i, and k is the number of equipment items. The manpower required to support equipment i is given by

$$m_i = T_0 N \beta \ , \tag{7.2}$$

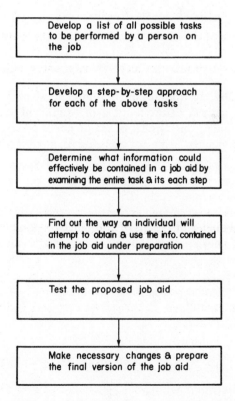

Figure 7.3. Steps of the procedure for developing job aids.

where N is the number of equipment items, T_0 is the equipment operating time (i.e., per day, per month, etc.), and β is the deficiency in man hr/equipment hr. The value of β is calculated from the following formula:

$$\beta = \frac{\theta_1}{\alpha_1} + \frac{c_1\theta_2}{\alpha_2} + \frac{c_2\theta_3}{\alpha_3} , \qquad (7.3)$$

where θ_1 is the mean net time to repair, in manhours; α_1 is the mean time to failure, in equipment hours; θ_2 is the mean net time to carry out all functions intended to stop failure, in manhours; α_2 is the equipment operating time needed for one preventive maintenance cycle, in equipment hours; θ_3 is the average net time to find out that the system is functioning adequately; α_3 is the equipment operating times between occasions for verifying that equipment is functioning adequately; and c_1, c_2 are constants.

For $c_1 = c_2 = 1$, Eq. (7.3) reduces to

$$\beta = \sum_{i=1}^{3} \frac{\theta_i}{\alpha_i} . \qquad (7.4)$$

Model II

This model can be used to find out the optimum crew size at a maintenance facility [15]. If we have θ_1 as the number of equipment items, which may need maintenance and θ_2 the number of maintenance persons, the probability, P_0, of no equipment waiting for or being repaired is given by

$$P_0 = 1 - \sum_{i=1}^{\theta_1} P_i , \qquad (7.5)$$

where P_i is the probability of i items of equipment waiting for maintenance or being repaired. For $0 \le i < \theta_2$, the value of P_i can be calculated from the following expression:

$$P_i = P_0 A_1 \left(\frac{\lambda}{r}\right)^i , \qquad (7.6)$$

where

$$A_1 \equiv \frac{\theta_1!}{i!(\theta_1 - i)!} , \qquad (7.7)$$

λ is the average failure rate of equipment, and r is the average rate of service.

Similarly, for $\theta_2 \le i \le \theta_i$, the value of P_i can be obtained from the following equation:

$$P_i = \frac{P_0 A_2 (\lambda/r)^i}{\theta_2^{(i-\theta_2)}} \, , \tag{7.8}$$

where

$$A_2 \equiv \frac{\theta_1!}{\theta_2!(\theta_1 - i)!} \, . \tag{7.9}$$

Substituting Eqs. (7.6) and (7.8) into Eq. (7.5) results in

$$P_0 = \left[\sum_{i=1}^{\theta_2-1} A_1 \left(\frac{\lambda}{r} \right)^i + \sum_{i=\theta_2}^{\theta_1} \frac{A_2(\lambda/r)^i}{\theta_2^{(i-\theta_2)}} \right]^{-1} \, . \tag{7.10}$$

Finally, the expected number of equipment items, N_e, waiting for maintenance or being repaired is given by

$$N_e = \sum_{i=0}^{\theta_1} i P_i \, . \tag{7.11}$$

The total cost, C_t, due to failure and repair is given by

$$C_t = k_1 + k_2$$
$$= N_e k_{md} + \theta_2 R_m \, , \tag{7.12}$$

where k_{md} is the cost per downtime hour per machine and R_m is the hourly cost of maintenance labor. The optimum value of θ_2 can be obtained from Eq. (7.12).

To find the optimum value of θ_2, the values of c_t have to be tabulated for $\theta_2 = 1, 2, 3, \ldots$. When the total cost is minimum, the value of θ_2 is at its optimum level.

THE MAINTENANCE REDUCTION CURVE

The human capacity for progressive learning of repetitive types of tasks is described by the learning curve. The learning curve concept is frequently used in manufacturing operations. Its validity in manufacturing operations has been verified by various researchers [16–18].

In maintenance work the learning curve is known as the maintenance reduction curve. In the past various researchers have applied it in maintenance work [19–22]. The maintenance reduction curve is quite useful for forecasting maintenance effort and manpower needs. It indicates the decreasing time needed to perform any repetitive maintenance operation as the operation is continued. Each time the maintenance operation is repeated, the curve forecasts the decrease in time. The decrease in time is the result of improved methods, procedures, work environment, worker familiarization and so on. The equation for the curve is given below:

$$T = \theta M^{-\beta} , \tag{7.13}$$

where β is the learning rate exponent, M is the number of maintenance tasks performed, T is the accumulated average time per maintenance task, and θ is the time needed to perform the first task.

The plot of Eq. (7.13) on log–log paper results in a straight line. A procedure to calculate the value of the exponent of the curve is given in Ref. [23].

SUMMARY

This chapter briefly presents the various aspects of human factors in maintenance and maintainability. The chapter begins by discussing the various aspects of human factors in maintenance. These are concerned with training and experience of maintenance personnel, diagnosis errors, maintenance data recording, equipment handling, and environmental factors.

Major tasks of maintainability engineering are presented along with elements of maintainability analysis. Useful guidelines, mostly in the form of questions, for professionals involved with maintenance and maintainability are discussed. These professionals are electrical designers, maintainability engineers, mechanical designers, maintenance technicians and maintenance engineers.

In addition, the chapter discusses recommendations for preparing maintenance procedures and a procedure for developing job aids. Two mathematical models associated with maintenance manpower requirements are presented. Finally, the learning curve concept is briefly described.

EXERCISES

1. List a number of recommendations for avoiding human errors in maintenance work.
2. What is the difference between the terms "maintenance engineering" and "maintainability engineering"?
3. What are the important elements of maintainability analysis?
4. What are the characteristics of a skilled maintenance technician?
5. Describe an approach used to develop job aids.

6. Discuss at least ten recommendations for developing maintenance procedures.
7. Discuss the history of the learning curve concept in maintenance work.

REFERENCES

1. Designing for maintainability, in *Human Engineering Guide to Equipment Design*, sponsored by Joint Army-Navy-Air Force Steering Committee. John Wiley & Sons, New York (1972).
2. J. D. Folley and J. W. Altman, *Guide to Design of Electronic Equipment for Maintainability*. Aerospace Medical Laboratory, Wright Air Development Center, Ohio (April 1956).
3. N. Jordon, *Human Factors Aspects in Maintainability*. Report No. P-2459, The RAND Corporation, Santa Monica, California (1961).
4. L. V. Rigby, *The Sandia Human Error Rate Bank*. Report No. SC-R-67-1150, Sandia Laboratories, Albuquerque, New Mexico (1967).
5. J. E. Robinson, W. E. Deutsch and J. G. Rogers, The field maintenance interface between human engineering and maintainability engineering. *Human Factors* **12**, 253–259 (1970).
6. M. Rapport and J. A. Cooper, *A Preliminary Study of the Human Factors Problems Associated with the Operation of the* [name is confidential] *Missile System*. Report No. ID-2274, Stanford Research Institute, California (November 1957).
7. J. M. Christensen and J. M. Howard, Field experience in maintenance, in *Human Detection and Diagnosis of System Failures* (Edited by J. Rasmussen and W. B. Rouse), pp. 111–133. Plenum Press, New York (1981).
8. D. Sauer, W. B. Campbell, N. R. Potter and W. B. Askren, *Relationships Between Human Resource Factors and Performance on Nuclear Missile Handling Tasks*. Report No. AFHRL-TR-76-85/AFWL-TR-76-301, Air Force Human Resources Laboratory/Air Force Weapons Laboratory, Wright-Patterson Air Force Base (Ohio 1976).
9. W. R. Downs, Maintainability analysis versus maintenance analysis—interfaces and discrimination, in *Proceedings of the Annual Reliability and Maintainability Symposium*, IEEE, New York, pp. 476–481 (1976).
10. W. Cox and C. E. Cunningham, *Applied Maintainability Engineering*, pp. 399–400. John Wiley & Sons, New York (1972).
11. A. D. Swain, *Maintenance Diagrams for Preventive Maintenance of Ground Electronic Equipment*. Report No. AIR-B20-57-FR-157 (October 1957). Available from American Institute for Research, Pittsburgh.
12. W. G. Ireson (Ed.), *Reliability Handbook*, pp. 11.2–11.2.1. McGraw-Hill, New York (1966).
13. J. D. Folley and S. J. Munger, *A Review of the Literature on Design of Informational Job Performance Aids*, Report No. ASD-TR-61-549 (October 1961). Available from Aerospace Medical Research Laboratories, Wright-Patterson Air Force Base, Ohio.
14. *Reliability of Military Electronic Equipment*, prepared by Advisory Group on Reliability of Electronic Equipment (AGREE) pp. 355–357. Available from the Superintendent of Documents, U.S. Government Printing Office, Washington, D.C. (June 1957).
15. R. Reed, *Plant Location, Layout, and Maintenance*. Irwin, Inc., Homewood, Illinois (1967).

16. A. Alchian, Reliability of progress curves in airframe production. *Econometrica* **31** (October 1964).
17. M. A. Rugero, An Economic Study of the Military Airframe Industry. Available from Wright–Patterson Air Force Base, Ohio (October 1957).
18. W. B. Hirschmann, Profit from the learning curve. *Harvard Business Review* **42**(4) (Oct. 1963).
19. M. A. Wilson, The learning curve in maintenance analysis, in *Proceedings of the Fifth Reliability and Maintainability Conference*, pp. 434–443 (1966). Available from the American Institute of Aeronautics and Astronautics, New York.
20. C. A. Bennett, Application of a learning curve to a maintenance problem, in *Proceedings of the Second Annual Quality Control Symposium*, Dallas (March 1957). Available from the American Society for Quality Control, Milwaukee, Wisconsin.
21. D. D. Gregor, Maintainability: F-5/T-38, design decisions, in *Proceedings of the Fourth Annual Reliability and Maintainability Conference*, pp. 509–525 (July 1965).
22. J. D. Patton, *Maintainability and Maintenance Management*, pp. 247–252. Instrument Society of America, Research Triangle Park, North Carolina (1980).
23. A. C. Laufer, *Operations Management*, pp. 274–280. South-Western Publishing Co., Cincinnati, Ohio (1975).

Chapter 8

Human Safety

INTRODUCTION

Human safety is an important factor in engineering systems design. However, prior to the end of the last century, the occurrences of industrial accidents were almost entirely blamed on workers. The employer shared very little responsibility for taking corrective measures in the workplace or on machinery or procedures [1]. Since World War II increasing emphasis has been placed on human safety. According to the beliefs of many safety experts, about 90% of accidents are related to unsafe acts, and only about 10% are due to unsafe physical or mechanical conditions. Some of the recent statistics associated with safety follow.

1. Over 10 million persons suffered a disabling injury [2], and 105,000 lost their lives in 1980 [2].
2. In 1980, the cost of accidents was over $83 billion.
3. Each year over 10,000 American workers lose their lives on the job [3].
4. In 1975, the government ordered recall of 200,000 trouble lights [3].
5. In 1969, about 35,700 transit bus passengers were injured in the United States [4]. Almost 16% of on-board accidents occurred in the acceleration mode, as opposed to 56% in the decelerating mode.
6. In 1962, there was only one case where the plaintiff was able to win a damage suit of $1 million or over, but in 1976, the number of cases were 43 [5].
7. More than 50% of the motor vehicles built in the United States were recalled for safety reasons since the inception of the motor vehicle recall program [5].
8. About 75% of the accidents occurred under conditions where a hazard was identified and corrective action taken [1].

The previous factors clearly demonstrate the necessity for human safety. Therefore, this chapter describes the various aspects of safety.

ACCIDENT LOSSES

The manufacturer of a product may be faced with various kinds of losses from an accident [5]. Examples of such losses follow.

1. Costs of damage suits and of settlements of death or injury claims.
2. Loss of prestige and public confidence.
3. Accident investigation cost.
4. Cost of preventive measures to avoid recurrence of accidents.
5. Decrease in morale.
6. Cost of increased insurance and of lost time of involved persons employed by the manufacturer.
7. Cost of slowdowns in operations during the accident-cause determination and during corrective measures.
8. Replacement cost of damaged equipment.
9. Loss of income from operations.
10. Cost of salvaging damaged equipment.

REASONS FOR ACCIDENTS

There are various reasons for the occurrence of accidents. The reasons and frequency of occurrence of accidents may very well vary from one segment of industry to another. For example, according to Ref. [6], the incident rate for injuries and illnesses for all manufacturing industry is 12.8 per 100 full-time workers per year as against 11.3 per 100 full-time workers per year for all mining industry. However, for all private industry the annual incident rate is 9.2 per 100 full-time workers.

Causes of Product Safety/Product Liability Problems

There are various causes which lead to product safety/product liability problems [7]. Some of them are as follows.

1. Wrong design and assembly.
2. Incorrect instructions and maintenance.
3. Poor training of operating and maintenance personnel.
4. Improper installation, handling and transportation.
5. Wrong testing.
6. Removal of safety devices.
7. Operator and maintainer errors.

According to the result of a study covering the period from July 1967 to March 1973, the fundamental causes of courtroom product liability cases were design defects, negligence in manufacturing, inadequate testing and inspection, incomplete instructions, defects in packaging, inadequate warning, incomplete instructions for use, installation and maintenance, and negligence in transportation.

Indirect Causes of Accidents

There are various proximal causes for the occurrence of accidents. Some of these are given below.

1. Mismatch between job and worker.
2. Inadequate worker attitude.
3. Departure from recommended safety standards concerning working conditions or work practices.
4. Inadequate management attitudes and skills.
5. Inadequately maintained tools and workplace.
6. Carelessly designed tools and workplace.
7. Failure of management to create suitable environments for an effective safety program.

Human Behavior

This section presents a number of typical human behaviors that may result in injury from an unsafe act [2]. Some of these behaviors are listed below.

1. Misreading instructions, scale markers and labels.
2. Failure to accurately estimate clearances or distances.
3. Failure to observe stated safety precautions.
4. Touching with hands to test.
5. Making use of a faulty product.
6. Taking for granted that operational or maintenance procedures are free of errors or omissions.
7. Failure to recognize the lack of adequate illumination.
8. Unwillingness to admit errors.
9. Failure to realize that overheated objects may lead to fire or explosion.
10. Failure to look into familiar surroundings when placing hands or feet.
11. Failure to respond rationally in emergency conditions.
12. Performing tasks without being mentally alert.

ACCIDENT- AND ERROR-REDUCTION MEASURES

This section discusses various accident- and error-reduction measures.

System Safety Functions

This section lists system safety functions. Effective performance of these functions will help to reduce the occurrences of accidents and human errors. Many of these functions are presented below [8].

1. Developing requirements to prevent the occurrence of accidents.
2. Participating in hazard analyses, design reviews, and accident investigations.
3. Developing plans for accident investigation.
4. Communicating information related to accident prevention.
5. Maintaining contacts with other safety organizations.
6. Maintaining information files concerning accidents/safety.
7. Evaluating emergency procedures.
8. Making recommendations for research and providing training for safety.
9. Negotiating and monitoring safety programs of suppliers.
10. Performing risk trade-off studies with respect to safety.

Measures for Preventing Accidental Injury

This section presents four basic measures in order of effectiveness [6].

1. Work toward eliminating the occurrence of hazard from the 3M's or P (i.e., material, method, machine or plant structure.)
2. Control hazards at their sources.
3. Provide training to concerned personnel to make them aware of hazards and follow hazard-avoidance procedures.
4. Provide protective equipment to concerned persons (equipment that will help to shield such persons against the hazard).

Measures for Improving Human Performance

Improved human performance will also help to reduce human errors and accidents [7]. Measures such as those that follow are useful for improving human performance.

1. Effectively motivate operators and management with respect to the prevention of accidents.
2. Provide training to operators and management in recognizing physiological and psychological stress, and conditions such as subordinate–superior relationships and family problems.
3. Effectively motivate design organization management with respect to safety.
4. Provide training to operators in job skills.
5. Avoid overloading operators.

6. Provide training to operators and management in controlling and reducing physiological and psychological stress.
7. Make use of job instructions and aids such as manuals, checklists and training films.

Useful Tools

The tools presented in this section can be used for various purposes; for example, improving operator performance, identifying errors, and reducing errors and their impact. Moreover, application of these tools will directly or indirectly influence the occurrence of accidents. Some of these tools follow [7].

1. Critical incident technique.
2. Techniques for stress modification and control.
3. Error-cause identification and removal system.
4. Human factors analysis, human reliability audit, job safety analysis and hazard analysis.
5. Physical and mental capability testing and psychological testing.
6. Quality control program and standardization.
7. Motivational methods and job instruction training program.
8. Reliability and maintainability analyses.
9. Failure mode and effects analysis (FMEA) and fault trees.

Safety Training

An effective training program ultimately will also help to reduce the occurrence of accidents. The steps shown in Fig. 8.1 are useful in developing a safety training program [2].

ACCIDENT INVESTIGATION

This aspect of human safety is as important as any other. Collection of inappropriate data and poor accident investigation will not help to reduce the occurrence of accidents. Equal importance must be given to this aspect of human safety because the information obtained through investigation and collected data will provide effective input to the accident reduction program.

Guidelines for Accident Investigators

Accident investigators should make use of guidelines such as the following when investigating accidents.

1. Make use of diagrams to identify the location of concerned personnel, equipment, etc. before and immediately after the occurrence of accidents.

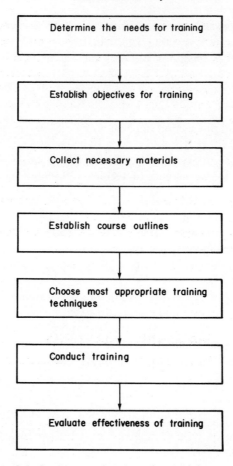

Figure 8.1. Seven steps for the safety training program.

2. Take photographs at the location of an accident whenever possible, from as many different angles as possible.
3. Take note of indoor conditions if the accident occurred indoors. Examples of indoor conditions are temperature, lighting, housekeeping, and warnings. However, if the accident occurred outdoors, take note of factors such as visibility and weather conditions.
4. Determine whether or not the employee was aware of the existence of unhealthy conditions in the case of an industrial accident.
5. Take proper care when obtaining statements from persons involved in accidents (eyewitnesses, etc.).

6. Make use of the fault tree method to develop potential causes if the accident cause is not readily clear.
7. Determine the extent of injuries received by the persons involved in accidents.
8. Find out if the item involved in the accident has been subjected to a failure analysis (the item whose failure caused the accident).

According to Ref. [2], the accident report should seek information on items such as those given below.

1. Nature of the injury and part of the body affected.
2. How the injury occurred and source of injury.
3. Kind of physical or environmental conditions that caused the hazardous event.
4. Kind of human action that caused the hazardous event.
5. Place of the accident.

Types of Accidents

Some of the basic accident types are given in Table 8.1.

MAN VERSUS MACHINES

This section discusses the strengths of humans and machines. According to Ref. [7] some of the areas in which man is better than machines are:

1. Inductive reasoning, tracking tasks and using judgments.
2. Performing under overloaded conditions.
3. Handling low-probability events and learning from past experiences.
4. Performance flexibility.
5. Reasoning out ambiguities and vague statements.

Table 8.1. Basic accident types

Number	Accident type
1	Caught in object or objects
2	Caught under object or objects
3	Struck against or fall from above
4	Rubbed or overexertion
5	Contact with electricity or toxic substances
6	Transit vehicle accidents
7	Exposure to radiation

6. Interpreting signals in the presence of distractions and high-noise environments.
7. Sensing abilities such as smell and taste.

Similarly, some of the areas in which machines are better than humans are computing, deductive reasoning, performing precise and repetitive jobs, handling sophisticated operations, response to control signals and insensitivity to emotions.

ACTIVITIES OF THE SAFETY ENGINEER AND SAFETY-WARNING-RELATED QUESTIONS FOR DESIGN ENGINEERS

Both the safety engineer and the designer play an important role in human safety. This section therefore presents the functions of a safety engineer and the safety-warning-related questions a design engineer seeks answers to during product design. Major activities associated with a typical safety engineer's job are shown in Fig. 8.2.

During the product design stage the product design engineer should ask various safety-warning-related questions to ensure safety [3].

1. Is there any need for special information for safely disassembling and repairing the product?

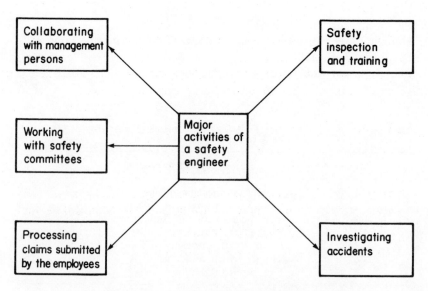

Figure 8.2. Activities of a safety engineer.

2. Is there any specific method that can shut down the product safely?
3. Is there any need for information to start the product safely?
4. Are there any specific maintenance needs that can help to enhance product safety?
5. Can special adjustments of the controls be helpful in fostering safe product operation?
6. Is there any recommended procedure for product installation?

SAFETY DEVICES AND HUMAN FAILURE MODES

This section addresses two important areas, i.e., safety devices and human failure modes. There are various kinds of devices used for safety purposes. The objective of these devices is to mitigate the impact of failures or accidents. Some of these devices follow [9]:

1. circuit breakers and fuses
2. helmets
3. photoelectric devices
4. key interlocks
5. seat belts
6. nets and air bags
7. pressure relief and limiting valves
8. limit switches
9. motion interlocks, ultrasonic and mercury switches
10. flame arrestors and fire suppressors
11. timers and explosion vents
12. electromagnetic sensing.

One or more of the human activities such as judgment, observation, action or lack of action, and decision to act can be associated with human error. Examples of human failure modes [10] are shown in Fig. 8.3.

FACTORS FOR POTENTIAL HUMAN ERRORS
AND ADVANTAGES OF ERROR STUDY
OVER ACCIDENT STUDY

This section briefly discusses both of these areas. In job safety analysis, consideration must be given to the potential for human error. Some of the factors for potential human errors follow.

1. Stress on the job.
2. Stress in private life.
3. Incorrect utilization of equipment.
4. Failure to minimize requirements for special employee training.

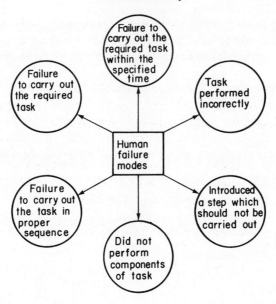

Figure 8.3. Human failure modes.

5. Worker tendency to take shortcuts in lengthy, unintelligible or uncomfortable procedures.
6. Difficulties in understanding written instructions and procedures.
7. Equipment that is difficult to maintain.

Errors are just like accidents. In many situations errors are the causes of accidents. To prevent the occurrence of errors there must be investigation. There are many advantages of error study over accident study. Some of these are listed below [11].

1. Error data are useful for providing clues to prevent the occurrence of accidents.
2. Error data are easier to obtain.
3. In comparison to accident descriptions, more objectivity is vested in error descriptions.
4. The utilization of error-reduction principles can be useful even in situations when no quantitative data are within reach.
5. Error analysis can enhance general performance.
6. More error data are available.
7. Some likely sources of trouble can be pinpointed even with fragmentary error data.

SELECTED SAFETY-RELATED FORMULAS

This section presents a number of formulas used in various safety studies.

Formula I

This formula is concerned with computing the value of the disabling-injury severity rate. This index is used to indicate the rate at which work days are charged or lost with respect to employee-hours of work. The index makes use of 1 million employee-hours of work [2]. The severity rate is given by

$$R_s = \frac{D_{tc}}{T} \times 1{,}000{,}000 \ , \tag{8.1}$$

where R_s is the severity rate, T is the employee-hours of exposure, and D_{tc} is the total days charged.

The value of the severity rate, say 250, represents that for each 1 million employee-hours worked the organization lost 250 days.

Formula II

This is used to calculate the value of the disabling-injury frequency rate. The frequency-rate index is defined below:

$$R_f = \frac{I_d}{T} \times 1{,}000{,}000 \ , \tag{8.2}$$

where R_f is the frequency rate and I_d is the number of disabling injuries. One should note here that this index is based on the factors given below for the specified period covered by the frequency rate.

1. Total number of deaths.
2. Total permanent and permanent partial disabilities.
3. Total temporary disabilities.

The numbers of the above injuries are expressed in terms of million-hour units by the index.

Formula III

This formula is concerned with computing the value of the accident rate in the aircraft industry [12]. The accident rate is defined as follows:

$$R_a = \frac{N_a}{H_f} \times 100{,}000 \ , \tag{8.3}$$

where R_a is the accident rate, N_a is the number of accidents, and H_f is the number of flight hours. Equation (8.3) gives the number of accidents per 100,000 flight hours.

Formula IV

This formula is used to calculate the value of risk. The formula is defined below [9]:

$$\alpha = \beta \sum_{k=1}^{\lambda} \sum_{i=1}^{\theta} D_k F_i C_{ki} , \qquad (8.4)$$

where θ is the number of hazard severity levels, λ is the number of damage states, β is the value at risk (this is defined in terms of dollars, number of people, etc.), F is the hazard occurrence probability, D is the expected fraction of damage from a certain hazard severity level, given that it happens, and C is the conditional probability that a certain damage state will happen, given that a certain severity of hazard has happened.

SUMMARY

This chapter presents various aspects of human safety. Various recent statistics associated with safety are first presented. Losses due to an accident are listed, along with the causes of the product safety/product liability problem. Indirect causes of accidents are also listed. Human behavior that may result in injury from an unsafe act is discussed.

Various accident- and error-reduction measures are presented; these are improving human performance, safety training, measures to prevent accidental injury, etc. System safety functions are listed along with tools to improve operator performance and identify and reduce errors. A number of guidelines for accident investigators are discussed along with basic accident types.

Strengths of man and machine are discussed. Activities of a safety engineer are outlined. A list of safety devices is presented. Factors for potential human errors are listed. Advantages of error study over accident study are briefly described. Finally, four safety-related formulas are presented. These formulas are concerned with computing severity rate, frequency rate, accident rate and risk.

EXERCISES

1. Describe at least eight examples of accident losses.
2. Discuss at least ten reasons for accidents.
3. What are the causes of the product liability problem?

4. Explain the important system safety functions.
5. What measures would you take to prevent accidental injury?
6. What are the measures to be taken to improve human performance?
7. What steps are necessary to develop a safety training program?
8. Discuss four basic accident types.

REFERENCES

1. T. A. Yoder, R. L. Lucas and G. D. Botzum, The marriage of human factors and safety in industry. *Human Factors* **15**, 197–205 (1973).
2. D. S. Gloss and M. G. Wardie, *Introduction to Safety Engineering*. John Wiley & Sons, New York (1984).
3. J. Kolb and S. S. Ross, *Product Safety and Liability*. McGraw-Hill, New York (1980).
4. J. A. Mateyka, Maintainability and safety of transit buses, in *Proceedings of the Annual Reliability and Maintainability Symposium*, IEEE, New York, pp. 1166–1225 (1974).
5. W. Hammer, *Product Safety Management and Engineering*. Prentice-Hall, Englewood Cliffs, NJ (1980).
6. J. M. Miller, The management of occupational safety, in *Handbook of Industrial Engineering* (Edited by G. Salvendy), pp. 6.14.1–6.14.18. John Wiley & Sons, New York (1982).
7. J. Mihalasky, The human factor in product safety, in *Proceedings of the Annual Conference of the American Society for Quality Control*, American Society for Quality Control, Milwaukee, Wisconsin, pp. 33–40 (1980).
8. J. A. Barton, Relationship and contributions of the system safety concept to cost and system effectiveness, in *Proceedings of the Fifth Reliability and Maintainability Conference*, pp. 501–504 (1966). Available from the American Institute of Aeronautics and Astronautics, New York.
9. J. Hrzina, Single-point failure analysis in system safety engineering. *Professional Safety* **16**, 20–26 (1980).
10. G. L. Wells, *Safety in Process Plant Design*, pp. 107–108. John Wiley & Sons, New York (1980).
11. W. G. Johnson, *MORT Safety Assurance Systems*. Marcel Dekker, New York (1980).
12. H. H. Moon and W. E. Knowles, Acceptance criteria for demonstrating system safety requirements, in *Proceedings of the Ninth Annual Reliability and Maintainability Conference*, IEEE, New York, pp. 440–444 (1970).

Chapter 9

Human Reliability Data

INTRODUCTION

Human reliability data play a vital role in reliability studies. In fact, human reliability data are the backbone of any human reliability study. Moreover, the accuracy of these studies will be directly affected by the quality of the input data. In other words, poor data will lead to misleading results. Therefore, as much attention as possible must be paid when collecting and analyzing reliability data. In addition, one has to be sure of what, when, why and where to collect human-reliability-related data because of the high price tag associated with the data collection.

At present, it may not be unreasonable to state that there is a lack of data in human reliability work in comparison to the availability of techniques and methods for predicting human reliability. Furthermore, the lack of human data is probably the most critical impeding factor to human performance reliability index development. There are various methods used to collect human-related data. Common standards such as reliability, validity, automation, economy, objectivity and quantifiability are to be applied to data-collection methods. This chapter discusses the various aspects of human reliability data.

GENERAL RELIABILITY DATA

Before specifically discussing human reliability data, it was considered necessary to familiarize readers with data-collection sources and failure reporting and documentation systems, because general knowledge of these areas will be useful for understanding the human-reliability-data collection problem. According to Ref. [1] there are nine identifiable data sources in the product life cycle. All of these must be carefully evaluated in data collection. Eight of the data-collection sources are found below.

1. *Past experience.* In this case, the data are collected from similar products used in the past.
2. *Customer's failure-reporting system.* Many users of engineering products have their own failure data banks. Therefore, this is a good source of obtaining failure data.
3. *Product development data.* This is another good data source. In this case data are obtained during the product development phase.
4. *Repair reports.* These are a good source of valuable data. These data are obtained from the repair facility.
5. *Field installation testing.* Field installation tests are a good source of data. These tests provide the first clear picture of equipment deficiencies in the field.
6. *Manufacturing and quality control.* This source generates an abundant quantity of data through inspection. Little effort is usually spent to obtain such data; however, they are the most contaminated.
7. *Acceptance tests in the factory.* These tests, conducted in the factory, also provide valuable data. These data are usually part of the manufacturing and quality-control documentation. However, it is possible to retrieve useful data from such documentation with minimal effort.
8. *Field demonstration and environmental qualification tests.* These tests are also a valuable source from which to collect data. In the product life cycle these are probably among the most carefully documented tests, because frequently these tests are involved in contract acceptance.

The following guidelines are useful in designing a failure reporting and documentation system [2].

1. Involve the end users of the system in design.
2. Keep the failure reporting form simple, clear and visible as much as possible.
3. Make the repair persons aware that documenting defects is as important as repairing the failed equipment.
4. Eliminate the need for memorizing codes.
5. Prerecord static information on the form whenever it is possible.
6. Analyze the recorded data as soon as possible and feed back the analyzed results to all concerned bodies.
7. Design a system that can also be used to specify manpower needs, initiate work requests, order spare parts, etc.

HUMAN FACTORS DATA

This section briefly discusses two aspects of human factors data. These are (1) human factors data in the system development phase, and (2) considerations and problems in using human factors data.

Human Factors Data in the System Development Phase

The system development phase may be subdivided into conceptual, definition and acquisition phases [3]. In the conceptual phase, providing mathematical human performance data is probably the most difficult problem. In this phase the data are required for function allocations, system effectiveness and availability evaluations.

Human factors data are also needed during the system definition phase. Here, human reliability data are required for inclusion in cases such as

1. human engineering maintainability design criteria,
2. system reliability calculations, and
3. time and performance information for maintainability.

Finally, in the acquisition phase the human components of the reliability and maintainability data are verified.

Considerations and Problems in Using Human Factors Data

In determining the applicability of potentially relevant human factors data to certain design problems, at least the following four factors must be considered [4].

1. Practical significance of the application.
2. Seriousness of the risks.
3. Caution in extrapolation of human data to other settings.
4. Trade-off considerations.

There are various impeding factors in applying human factors data. According to the findings of some experts, many design engineers neglect to pay attention to human factors in their design approaches. Some of the possible reasons for this problem may be as follows.

1. Poor quality of the human factors design standards in engineering design.
2. Faulty education of design engineers.
3. Resistance from design engineers themselves.
4. Resistance from managers.

5. Lack of education of human factors specialists.
6. Lack of communication.
7. Presentation of human factors data in an inadequate form.

GUIDELINES FOR HUMAN PERFORMANCE RELIABILITY DATA SYSTEM DEVELOPMENT

Some of these guidelines follow [5,6].

1. Ensure that the data retrieval process is short and simple.
2. Ensure that the data system is flexible so that it can accept data from a variety of sources.
3. Ensure that the definitions and terms used are meaningful to system users.
4. Ensure that the data system has the statistical analysis capability to analyze its own data.
5. Ensure that a significant part of the human performance reliability data can be integrated with product reliability predictive data.
6. Ensure that the performance data can be associated with various combinations of man–machine components.
7. Ensure that the data system is compatible with user circumstances.
8. Ensure that the output of the data system is in an effective format.

HUMAN PERFORMANCE DATA COLLECTION METHODS

There are various methods used to collect human data. Six common standards shown in Fig. 9.1 are applied to data collection techniques [7]. Data collection techniques can be grouped into the following four categories:

Category I. This category includes the direct manual methods. These methods are composed of time and motion study techniques, continuous observations, demonstrations and sampled observations.
Category II. This category includes all of the indirect manual methods. Methods such as proficiency tests, interviews and ratings, problem-incident reports and questionnaires fall under this category.
Category III. This category includes system measurement records. Examples of such records are maintenance records, system test records and human-initiated failures.
Category IV. This category includes all of the automatic methods. Examples of such methods are physiological-response recording instrumentation and instrumentation for task performance recording.

Automatic methods are used in situations such as the following.

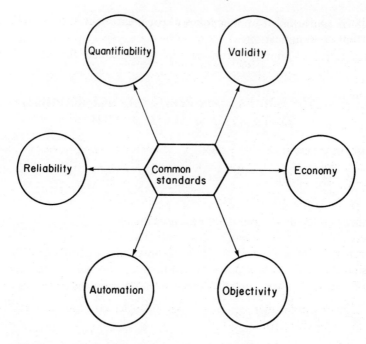

Figure 9.1. Common standards applied to data collection techniques.

(a) Inadequate workspace for data collectors.
(b) System outputs are in a large number.
(c) Difficult to sense system outputs directly.
(d) Hazardous environments for data collectors.

Benefits and Drawbacks of Human Data Collection Methods

There are various advantages and disadvantages of automatic and manual recording methods. Some of the advantages of instrumentation recorders follow.

1. Eliminates the possibility of human error.
2. Accuracy and sensitivity are better than other methods in many applications.
3. On-line data recording is rapid.
4. Eliminates the possibility of human bias in data.
5. Permits measurements in situations where humans have difficulties.

Similarly, some of the disadvantages of the instrumentation recorder are listed below.

1. Maintenance is required on a continuous basis.
2. There is a possibility of catastrophic failures.
3. It could be quite costly.
4. Transportation and installation could be quite difficult.
5. Frequently it is too specialized.

On the other hand, there are also various benefits and drawbacks to the use of a human recorder for data collection. Some of the benefits of the human recorder are as follows.

1. The requirement for maintenance is either very little or nil.
2. Chances for a catastrophic failure are negligible.
3. Relatively mobile and compact.
4. Possess the ability to carry out certain evaluations in situations where a machine is unable to do so.
5. The flexibility to adapt and the ability to interpret during recording exist.

Just as in any other method, the human recorder also has various drawbacks. Some of these follow [7].

1. Frequently less sensitive in comparison to machines.
2. The possibility of occurrence of human error exists.
3. Normally less efficient in data recording relative to instrumentation.
4. The possibility of bias in data recording exists.

DATA BANKS FOR HUMAN RELIABILITY STUDIES

As mentioned earlier, data banks play an important role in human reliability analysis. In fact, they may be called the back bones of reliability studies because poor-quality input data will result in misleading conclusions from reliability analysis.

Even today, the poor availability of good human reliability data is still a vital problem to be overcome. Many experts have proposed various kinds of formats for human reliability data banks [8–11]. However, according to Ref. [10] an ideal data bank in human reliability work should provide data relating to factors such as those given in Table 9.1. However, the existence of an ideal human unreliability data bank is an impossibility. At present, it is within reach to have at least the data bank formats shown in Fig. 9.2.

The next question that arises is what are the possibilities of securing human performance reliability data? The answer to this question is that there are three ways to obtain such data. These are experts' judgments, field operations

Table 9.1. Some of the data-relating factors for an ideal data bank

Factor No.	Factor description
1	Types of tasks
2	Environmental conditions
3	System elements
4	System characteristics
5	Types of systems
6	Motivation
7	Skill and training of concerned persons
8	Psychological stress
9	Quality of written instructions

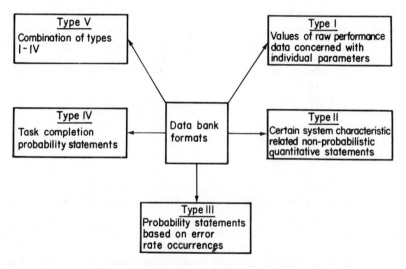

Figure 9.2. Possible data bank formats.

and laboratory studies. Thus, three different types of human performance data banks can be developed. Each of these data banks is described below.

Type I Data Banks

The basis for input data for these data banks are the judgments of experts. Researchers such as those listed in Table 9.2 [12–16] have contributed to subjective-based data banks. There is a wide variation in techniques for devel-

Table 9.2. Subjective data bank contributors

No.	Researchers	Brief description of contribution	Reference number in this chapter
1	L. W. Rook	Application of subjective rankings to get estimates for human performance reliability	[12]
2	W. B. Askren	Application of scales to establish subjective estimates of the relationship between performance of operators and personnel factors	[13]
3	N. Dalkey, F. Helmer	Experimental application of the DELPHI technique to the use of experts	[14]
4	R. L. Smith	Technique for establishing personnel performance standards	[15]
5	A. I. Siegel, P. J. Federman	Derivation of human performance reliability indices from critical incidents of extremely effective and ineffective performances	[16]

oping expert estimates. They vary from less-rigorous to very-rigorous psychophysical methods. An example of very rigorous techniques is the one developed by Siegel in Ref. [16]. Similarly, the DELPHI technique is a prime example of the less-rigorous method. This method narrows the judgmental variations of the judges by feeding back the overall results to each of these experts. This causes them to revise their earlier estimates until a consensus is reached. This technique was used at the United States Navy Personnel Research and Development Center in San Diego, California [17]. Subjective data banks can be quite useful provided that factors such as those given below are considered:

1. selection of judges
2. validity of data
3. performance boundaries
4. judgment description level
5. estimation procedure.

When selecting judges for obtaining subjective task data, attention must be paid to the fact that only those people are selected who have extensive experience in performing the tasks in question and who have observed others performing such tasks. With respect to the validity of data, the data obtained from expert opinions will contain some error. Therefore, consideration must be given to the potential usefulness of data. However, with effective data collection procedures, the magnitudes of errors can be made very small. The third factor listed previously is concerned with performance boundaries. In this case, the boundaries of the performance being measured must be effectively evaluated at an early stage. The judgment description level and performance-shaping factors associated with estimates must also be evaluated at an advanced stage. In addition, identification of the types of errors to be considered is required. Finally, the last factor is concerned with specifying a procedure for obtaining estimates. Examples of such estimating procedures are Thurstone's paired comparisons and DELPHI.

The following are some of the advantages and disadvantages of the subjective human performance data bank.

Advantages.

1. It is relatively less expensive to develop, and will contain the range of parameters for which a data bank is required.
2. It is relatively easy to establish because a large amount of data can be obtained from a small number of judges.

Disadvantages.

1. The validity of the data will be unknown in the initial stage.
2. It may not be comparatively that precise.
3. It will contain some error.

Type II Data Banks

Laboratory sources provide input data to these banks. One can have utmost confidence in such data, because these are least influenced by subjective elements, and of course subjective elements generate error. One well-known data bank, known as the Data Store [18], is based on laboratory sources. This bank is based on 164 studies, and explains several characteristics of commonly used displays and controls. All of these studies were extracted from several thousand reports.

One of the important shortcomings of the Data Store is that it does not take into consideration many of the performance shaping factors.

Considering the case of time-continuous tasks, human performance reliability data were collected and analysed in Ref. [19]. These data were collected from an experiment concerning a vigilance task. In this experiment, the subjects were asked to observe a clock-type display of lights. In addition, when a failed-light event occurred the subject responded to it by pressing a hand-held switch. Fifty one male and female university-level students and United States Air Force personnel served as subjects for the study.

Two types of data were gathered from this study: miss error and false-alarm error. The miss error is associated with a situation where the subject failed to detect the failed light. On the other hand, the false-alarm error is associated with a situation where the subject acted if a failed-light event had occurred when in fact it did not.

Type III Data Banks

These data banks are known as field-based data banks. Input data to these data banks come from operational sources. These kinds of data are more realistic than those of previous kinds but are more difficult to obtain. There are two noteworthy data banks [20,21]; these are the Operational Performance Recording and Evaluation Data System (OPREDS) and the Sandia Human Error Rate Bank (SHERB). The OPREDS data bank is the result of the efforts of researchers at the United States Naval Ocean System Center. The system developed by these researchers allowed computer monitoring of operator actions at 15 Navy Tactical Data System Stations automatically. Operator actions are concerned with switch actions. The main drawback of this system is that it is limited by its ability to deal with restricted operations only, such as switch actions.

The second data bank, known as the Sandia Human Error Rate Bank, was proposed by researchers at Sandia Laboratories [10,21]. This data bank obtains input data from commercial–industrial and military sources.

SOURCES OF HUMAN-RELATED DATA
IN PROCESS PLANTS

In order to perform various types of human-reliability-related analysis in process plants, the data can be gathered from several different sources. Examples of the data that can be collected are [22]

1. human errors
2. number of accidents
3. nature of accidents and major incidents
4. number of major incidents.

Information sources for the number and nature of accidents are training personnel, plant designers, safety personnel and plant managers. Similarly, the information sources for the number and nature of major incidents are insurance companies, plant managers and safety personnel. Finally, human error information can be obtained from people such as plant managers, safety personnel, plant designers, training personnel and control system designers.

Human Error Classifications in Process Plants

For data collection purposes, human errors can be classified into categories such as isolated acts, vigilance tasks (e.g., critical alarm signal undetected), simple tasks, emergency responses, complex tasks, control tasks (e.g., critical control error committed) and operator incapacities. The principal information associated with each of these categories is the number of acts per unit

Table 9.3. Human error rates for selected tasks

Number	Error description	Error rate per million operations	Error rate* in errors/ plant-month
1	Reading gauge incorrectly	5000	—
2	Closing valve improperly	1800	—
3	Soldering connectors improperly	6460	—
4	Actuating switch inappropriately (observe chart)†	1128	—
5	Failure to tighten nut and bolt	4800	—
6	Failure to install nut and bolt	600	—
7	Improper adjustment of mechanical linkage	16700	—
8	Procedural error in reading instructions	64500	—
9	Connecting hose improperly	4700	—
10	Failure to pursue proper procedure by the operator	—	0.0401
11	Improper adjustment by the maintenance personnel	—	0.0134
12	Installation error	—	0.0401
13	Misinterpretation or misunderstanding of requirements by the operator	—	0.0076
14	Inadvertent or improper equipment manipulation by the operator	—	0.0706
15	Improper servicing or reassembly by the maintenance personnel	—	0.0153

*These error rates are associated with pressurized water reactors (PWRs).
†Description is given in parentheses.

Table 9.4. Selected human reliability data sources

Number	Author(s)	Title	Published in or prepared by
1	J. L. Recht	Systems safety analysis: Error rates and costs	*National Safety News*, pp. 20–23 (Feb. 1966)
2	D. Meister	Human factors in reliability	*Reliability Handbook* (Edited by W. G. Ireson), pp. 12.2–12.37. McGraw-Hill, New York (1966)
3	G. A. Peters	Human error: Analysis and control	*American Society of Safety Engineers*, **XI**, 9–15 (1966)
4	A. D. Swain, H. E. Guttmann	*Handbook of Human Reliability Analysis with Emphasis on Nuclear Power Plant Applications*	Report NUREG/CR-1278 (draft), United States Nuclear Regulatory Commission, Washington, D.C. (1983)
5	D. W. Joos, Z. A. Sabri, A. A. Husseiny	Analysis of gross error rates in the operation of commercial nuclear power stations	*Nuclear Engineering and Design*, **52**, 265–300 (1979)
6	T. A. Kletz	The uses, availability and pitfalls of data on reliability	*Process Technology International*, **18**, 111–113 (1973)
7	E. W. Hagen	Human reliability analysis	*Nuclear Safety*, **17**, 315–326 (1976)
8	T. A. Kletz, G. D. Whitaker	*Human Error and Plant Operation*	Report EDN-4099, Safety and Loss Prevention Group, Petrochemicals Division, Imperial Chemical Industries, Ltd., Billingham, England (1973)
9	T. A. Regulinski W. B. Askren	Mathematical modeling of human performance reliability	*Proceedings of the Annual Symposium on Reliability*, 1969, pp. 5–11. Available from the Institute of Electrical and Electronic Engineers, New York
10	G. Berry	The Weibull distribution as a human performance descriptor	*IEEE Transactions on Systems, Man, and Cybernetics*, **11**, 501–504 (1981)

continued

Table 9.4. (continued)

Number	Author(s)	Title	Published in or prepared by
11	G. M. Hunns	Discussions around a human factors data-base	*High Risk Safety Technology* (Edited by A. E. Green). John Wiley & Sons, Chichester (1982)
12	J. F. Ablitt	*A Quantitative Approach to the Evaluation of the Safety Function of Operators on Nuclear Reactors*	Report AHSB (s)R-160, United Kingdom Atomic Energy Authority (UKAEA), Warrington, U.K. (1969)
13	S. J. Munger	*An Index of Electronic Equipment Operability: Data Store*	Report AIR-C43-1/62-RP(1), American Institute for Research, Pittsburgh, PA (1962)
14	R. L. Smith	*Technique for Establishing Personnel Performance Standards*	Personnel Research Division, Report No. PTB-70-5, Bureau of Naval Personnel, Washington, D.C. (December 1969)
15	R. Urmston	*Operational Performance Recording and Evaluation Data System*	Navy Electronics Laboratory Center, San Diego, CA (November 1971)
16	H. I. Williams	Reliability evaluation of the human component in man-machine systems	*Electrical Manufacturing*, **4**, 78–82 (1958)
17	—	*Reactor Safety Study — An Assessment of Accident Risks in U.S. Commercial Nuclear Power Plants*	U.S. Nuclear Regulatory Commission Report WASH-1400 (NUREG-75/014), Washington, D.C. (October, 1975)
18	—	*Anthropometric Source Book*, Vols. 1–3	NASA. Available from NASA Reference Publication 1024, Lyndon B. Johnson Space Center, Houston, Texas

continued

Table 9.4. (continued)

Number	Author(s)	Title	Published in or prepared by
19	R. P. Kern*	0025 *Data Base*	U.S. Army. Available from PERI-II, Army Research Inst., 5001 Eisenhower Avenue, Alexandria, VA 22333
20	A. I. Siegel*	APS data	Applied Psychological Services (APS), 404 E. Lancaster, Wayne, PA 19087
21	H. L. Parris*	*Anthropometry*	Electric Power Research Inst., P.O. Box 10412, Palo Alto, CA 94303
22	P. M. Haas*	*Safety Related Operator Actions (SROA)*	Oak Ridge National Laboratory, Bldg. 6025, P.O. Box X, Oak Ridge, TN 37830
23	Wm. Hannaman*	*Gas Cooled Reactor (GRC) Data Base*	General Atomic, P.O. Box 81608, San Diego, CA 92131
24	W. W. Banks	*Dual Scale Meters: Visual Reading Errors*	EG&G Idaho Inc., P.O. Box 1625, Idaho Falls, ID 83415

*Contact person.

time, the number of failures per unit time, the number of task demands and of failures, the number of emergencies and of ineffective behaviors per emergency, the number of failures per unit time, and the number of incapacities per unit time.

HUMAN ERROR DATA FOR SELECTED TASKS AND HUMAN RELIABILITY DATA SOURCES

The objective of this section is to present human error data for selective tasks and selective human reliability data sources.

The human error data for selective tasks are taken from Refs. [23-25]. These data are presented in Table 9.3. Selective human reliability data sources are given in Table 9.4. These sources present human reliability data on various areas.

SUMMARY

This chapter briefly discusses the various aspects of human reliability data. Eight general reliability data sources are described. A number of guidelines for designing a failure reporting and documentation system are presented. Human factors data in the system development phase are discussed, along with considerations and problems of using such data. Various guidelines for human performance reliability data system development are outlined. Four classes of human performance data collection methods are briefly discussed. Advantages and disadvantages of automatic and manual recording methods are presented. Three types of data banks for human reliability studies are described; these are experimentally based data banks, subjective-based data banks and operationally based data banks. Some of the data-relating factors for an ideal data bank are tabulated. Five different types of data bank formats are outlined. Sources of human-related data in process plants are discussed, along with human error classifications in process plants.

Human error data for various selective tasks are tabulated. Seventeen sources for human reliability data are given in a tabular form.

EXERCISES

1. What are the main sources for general reliability data?
2. Describe at least seven guidelines useful in designing a failure reporting and documentation system.
3. What are the difficulties in using human factors data?
4. What are the advantages and disadvantages of automatic human performance data collection methods?
5. Discuss the following data banks:
   a. data store
   b. Sandia human error rate bank
   c. operational performance recording and evaluation data system.
6. Describe the human error classifications applicable in process plants.
7. Discuss the characteristics of an ideal human error data bank.
8. Discuss the current developments in human reliability data collection.

REFERENCES

1. R. F. Hahn, Data collection techniques, in *Proceedings of the Annual Reliability and Maintainability Symposium*, IEEE, New York, pp. 38–43 (1972).
2. E. T. Parascos, A new approach to the establishment and maintenance of equipment failure rate data bases, in *Failure Prevention and Reliability* (Edited by S. B. Bennett, A. L. Ross and P. Z. Zemanick), pp. 263–268. American Society of Mechanical Engineers, New York (1977).
3. D. B. Jones, The need for quantification in human factors engineering, in *Pro-

ceedings of the Sixth Annual Reliability and Maintainability Conference, American Society of Mechanical Engineers, New York, pp. 87–92 (1967).

4. E. J. McCormick and M. S. Sanders, *Human Factors in Engineering and Design*, pp. 486–492. McGraw-Hill, New York (1982).
5. R. E. Blanchard, Human performance and personnel resource data store design guidelines. *Human Factors* **17**, 25–34 (1975).
6. D. Meister and R. G. Mills, Development of a human performance reliability data system, in *Proceedings of the Annual Reliability and Maintainability Conference*, American Society of Mechanical Engineers, New York, pp. 425–439 (1972).
7. D. Meister, G. F. Rabideau, *Human Factors Evaluation in System Development*. John Wiley & Sons, New York (1965).
8. A. D. Swain, Development of a human error rate data bank, in *Proceedings of the U.S. Navy Human Reliability Workshop*, NAVSHIPS 0967-412-4010. Available from the Naval Publications and Forms Center, 5801 Tabor Avenue, Philadelphia, PA 19120 (February 1971).
9. R. E. Blanchard, Human performance and personnel resource data store design guidelines. *Human Factors* **17**, 25–34 (1975).
10. D. Meister, Subjective data in human reliability estimates, in *Proceedings of the Annual Reliability and Maintainability Symposium*, IEEE, New York, pp. 380–384 (1978).
11. B. S. Dhillon and C. Singh, *Engineering Reliability: New Techniques and Applications*, Chap. 7. John Wiley & Sons, New York (1981).
12. L. W. Rook, Evaluation of system performance from rank-order data. *Human Factors* **6**, 533–536 (1964).
13. W. B. Askren, *Feasibility of a Computer Simulation Method for Evaluating Human Effects on Nuclear System Safety*. Report No. AFHRL-TR-76-18, Air Force Human Resources Laboratory, Brooks Air Force Base, Texas (May 1976).
14. N. Dalkey and F. Helmer, An experimental application of the DELPHI method to the use of experts. *Management Sciences* **10**, 458–467 (1963).
15. R. L. Smith, *Technique for Establishing Personnel Performance Standards* (TEPPS). Report No. PTB-70-5, Vols. I–III, Personnel Research Division, Bureau of Naval Personnel, Washington, D.C. (December 1969).
16. A. I. Siegel and P. J. Federman, *Investigation into and application of a fleet post-training performance evaluation system*. Report No. 7071-2, available from Applied Psychological Services, Inc., Wayne, Pennsylvania (September, 1970).
17. O. A. Larsen and S. I. Sander, *Development of Unit Performance Effectiveness Measures Using DELPHI Procedures*. Report No. NPRDC TR-76-12, Navy Personnel Research and Development Center, San Diego, CA (September 1975).
18. S. J. Munger, *An Index of Electronic Equipment Operability: Data Store*. Report No. AIR-C43-1/62-RP(1), Prepared by the American Institute for Research, Pittsburgh, PA (January 1962).
19. T. L. Regulinski and W. B. Askren, Mathematical modeling of human performance reliability, in *Proceedings of the Annual Symposium on Reliability*. Available from IEEE, New York, pp. 5–11 (1969).
20. R. Urmston, *Operational Performance Recording and Evaluation Data System*, Developed by Navy Electronics Laboratory Center, San Diego, California (November, 1971).
21. A. D. Swain, Development of a human error rate data bank, in *Proceedings of the U.S. Navy Human Reliability Workshop*, Report No. NAVSHIPS 0967-412-4010 (Feb. 1971). Available from the Naval Publications and Forms Center, 5801 Tabor Avenue, Philadelphia, PA 19120.

22. F. P. Lees, Quantification of man-machine system reliability in process control. *IEEE Transactions on Reliability* **22**, 124-131 (1973).
23. J. L. Recht, Systems safety analysis: Error rates and costs. *National Safety News*, 20-23 (February 1966).
24. G. A. Peters, Human error: Analysis and control. *American Society of Safety Engineers Journal* **XI**, 9-15 (1966).
25. D. W. Joos, Z. A. Sabri and A. A. Husseiny, Analysis of gross error rates in operation of commercial nuclear power stations. *Nuclear Engineering and Design* **52**, 265-300 (1979).

Chapter 10

Human Factors in Quality Control

INTRODUCTION

Human factors is one of the newer engineering disciplines and plays an important role in quality control. Many quality problems are associated with humans. For example, according to Ref. [1] over 50% of all quality problems are due to human mistakes. Furthermore, according to a study of nuclear weapon systems, 82% of defects were due to human errors.

In a quality assurance system people perform various functions. Several types of human errors are associated with each task. Some examples of human functions in quality assurance are [2]

1. testing
2. inspection
3. selection of materials
4. design of processes
5. assembly.

A person performing a quality function is influenced by organizational, physical and individual factors.

The followings items are included in the organizational, physical and individual factors.

Organizational factors

1. supervisory practices
2. work methods and procedures
3. policies and work group type
4. social aspects of the organization.

Physical factors

1. tools and aids
2. equipment
3. layout of work place.

Individual factors

1. interests and attitude
2. knowledge and skill
3. temperament.

This chapter presents the various aspects of human factors in quality control.

HUMAN ELEMENT CONSIDERATIONS
IN QUALITY ASSURANCE

According to Ref. [2], human-related considerations such as those that follow are to be considered in quality assurance.

1. Selection of competent persons for performing quality work.
2. Selection and development of equipment, instruments and tools useful to persons selected to carry out high-quality work.
3. Communication of quality specifications to concerned persons.
4. Finding ways for matching quality goals with the personal goals of individuals.
5. Creating quality consciousness within persons.
6. Providing information for quality-related decisions.
7. Developing procedures to obtain quality-related information.
8. Organizing quality tasks into jobs in such a way that they match effectively with the skills and knowledge of involved persons.

MANAGEMENT- AND OPERATOR-CONTROLLABLE
ERRORS IN QUALITY CONTROL

This section briefly discusses the two types of errors in quality control. These are management-controllable errors and operator-controllable errors.

According to Refs. [3,4] management-controllable errors may be as high as about 80% of total errors. In order to reduce such errors, management plays an important role by assuring that the operator or worker is in a state of self-control. The following are necessary for the worker or operator to be in a state of self-control:

1. knowledge of what he or she is actually doing;
2. knowledge of what he or she is expected to do;
3. possession of the means to regulate things when a correction is necessary.

Finally, it may be said that if any of the criteria for self-control have not been satisfied, then a defect is management controllable. Some examples are improper tools, defective machinery, poor accuracy of measuring instruments, wrong work instructions and blueprints, poor training, inadequate lighting and crowded working area.

On the other hand, when all of the criteria for self-control have been provided, a defect is said to be operator controllable. The operator or worker is responsible for errors occurring under such conditions.

FUNCTIONS OF QUALITY SUPERVISORS

The role of supervisors is important for the effective functioning of the quality system. The results of poor supervision will be reflected in the poor performance of quality personnel. In many instances the cause of ineffective supervision is due to the supervisor not being clearly aware of his or her functions.

All quality supervisors are expected to perform certain functions. These functions may be classified into the five categories shown in Fig. 10.1. The diagram shows the following five functions:

1. developing plans
2. making decisions
3. communication
4. monitoring and assessing performance
5. personnel development.

Developing plans is a vital function of any quality supervisor. Success or failure of any quality operation depends on plans formulated by a supervisor for future needs. Quality supervisors must pay attention to planning activities such as those below.

1. Determining and forecasting the need for quality-related information for future use.
2. Determining and forecasting the need for inspection manpower for future quality programs.
3. Determining and forecasting the need for a new product with respect to inspection equipment and task requirements.
4. Determining and scheduling the training needs of quality personnel for future programs.
5. Setting both short-term and long-term goals with persons involved in the quality function.

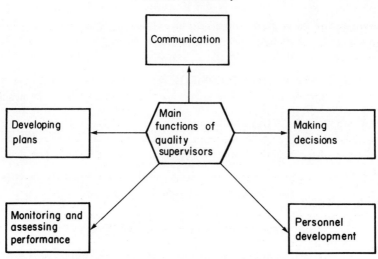

Figure 10.1. Main functions of quality supervisors.

Decision making is another function performed by quality supervisors. In broader terms, it involves collecting information, organizing it into necessary classifications and choosing an action. In the quality system, decisions usually involve management of resources. Resources are composed of men, machines and information.

The next function is concerned with communication. Poor communication between supervisors and workers will lead to unhealthy results. It is important that quality supervisors understand that communication takes place only when a sound understanding is reached which results in a desirable action. Oral and written communications are the two approaches used for transferring information. Each of these methods has certain advantages. Some of the advantages of oral communications are (i) personal effectiveness, (ii) exchange of information, and (iii) flexibility. Similarly, the main advantages of written communications are (i) permanent documentation for future use and (ii) accuracy and authority. Quality supervisors are required to have command of both these methods.

Monitoring and assessing performance is another function performed by the quality supervisors. Here supervisors are required to monitor and assess the performance of their workers. To perform this function effectively, the supervisor must make use of correct and accurate monitoring and assessing approaches.

Finally, personnel development is another function performed by the quality supervisors. This function includes expanding the employee's capabilities by providing opportunities for learning new tasks, identifying and training employees with high potential for management work, etc.

INSPECTION

Inspection is a vital aspect of the quality control program. Furthermore, the human element plays an important role in inspection. These are various purposes of inspection [5]. Some of them are shown in Fig. 10.2. To accomplish any of these objectives the human element must be involved to a certain degree.

Inspection Tasks

This section briefly examines the basic components of inspection tasks and the types of inspection tasks. The basic components of inspection tasks are shown in Fig. 10.3. The diagram in this figure shows four basic components. These are interpretation, comparison, decision making and action. Industrial inspection is conducted with the aid of some sort of established standards.

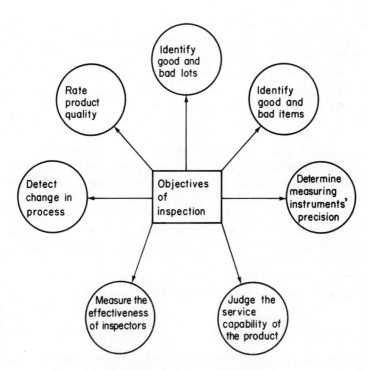

Figure 10.2. Objectives of inspection.

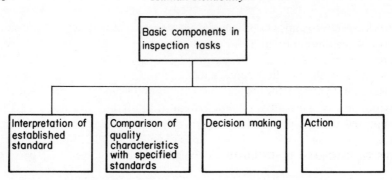

Figure 10.3. Basic components in inspection tasks.

These standards define what is acceptable and what is not. Therefore, the correct interpretation of these standards is necessary for effective inspection. The next component of inspection tasks is concerned with comparison. Here the inspector compares the quality characteristics with defined standards. The decision-making component involves decisions on whether or not the quality characteristic conforms to the specified standard. Finally, the action element of inspection tasks is concerned with actions to be taken by the inspector on the basis of his or her decisions.

Inspection tasks may be grouped into three basic classes. These are

1. measurement tasks
2. scanning tasks
3. monitoring tasks.

The first kind of tasks, namely measurement tasks, include those inspections in which dimensions of items are measured with the aid of measuring instruments to determine if the specified limits are satisfied. Scanning tasks involve point-by-point examination of an item by the inspector to discover defects. Scanning tasks are probably the most common tasks. Monitoring tasks are concerned with a situation such as control of some kind of automatic system or equipment. In this situation the displays are monitored by the inspector for out-of-tolerance conditions.

Factors that Influence the Accuracy of Inspection

The accuracy of inspection is influenced by various factors. Some of these factors follow [3].

1. Complexity and sophistication of the product. These products are difficult to inspect.

2. Poor eyesight of inspectors and fatigue.
3. Product orientation.
4. Presence of people.
5. Variation in the product defect rate.
6. Inspection hypnosis (this means that there is an inclination to see what one expects to see when inspecting a familiar product).
7. Repeated inspections.

Ways to Improve Inspection Performance

To improve inspection performance it is advisable to pay attention to the following items.

Layout of inspection stations. This influences the inspection performance. A properly designed inspection station will be helpful for improving inspection performance. When designing an inspection station consideration must be given to factors such as lighting, seating, and access to incoming and outgoing materials.

Written inspection instructions. Good written instructions are vital to inspectors for performing their function effectively. One or more of the six basic types of information shown in Fig. 10.4 should be included in all written instructions. The following guidelines are useful for writing inspection instructions.

1. Avoid using adjectives and adverbs.
2. Establish a clear-cut objective for each instruction.

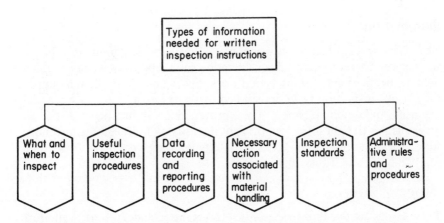

Figure 10.4. Types of information needed for written inspection instructions.

3. Use commonly used words.
4. State the correct inspection method in operational terms.
5. Use only necessary words.

Training. Training should be given to inspectors whenever it is necessary, and learning should be a continuous process. Fast changes in technology affect the quality function. Training may take various forms, e.g., courses in local technical institutes, short courses from professional societies, or self-instruction from books and manuals.

Tools and methods. These affect the performance of inspectors. Therefore, every effort is to be made to provide inspectors with effective tools and methods.

Quality standards. Good quality standards help to improve inspection performance. Every effort should be made to make use of up-to-date quality standards and to keep the number of different standards used by an inspector to a minimum level.

Repeated inspections. To improve inspection accuracy repeated inspections are another useful tool. However, it is a costly one. Therefore, it should be restricted to critical products only.

Inspection search approaches. Two procedures are followed to discover defects in a product. The first one is concerned with searching one area for all characteristics and then going over another. On the other hand, the second procedure calls for examining one characteristic at a time over the entire product. Past experience indicates that the second procedure is more efficient than the first one.

Illumination

The effectiveness of visual inspection is subject to the illumination in the workplace. Therefore, special consideration should be given when designing lighting arrangements for inspection work. Special needs for illumination of the various inspection tasks are studied by professionals such as industrial engineers and quality planners. These professionals should pay attention to areas such as (i) light diffusion, (ii) light intensity and (iii) lighting direction. Recommended illumination levels for selected purposes are taken from Refs. [2,4] and given in Table 10.1.

Inspector Errors

These errors are due to inspectors. Inspectors miss about 20% of the defects in products [4]. Their errors may be classified into the classifications shown in Fig. 10.5.

Table 10.1 Recommended levels of illumination for various
types of inspection work

Item No.	Type of work	Light intensity (foot candles)
1	Microscopic examination of materials	500
2	Ordinary inspection	50
3	Unmagnified visual, functional and dimensional product inspection task	100
4	Difficult inspection task	100
5	Highly difficult inspection	200

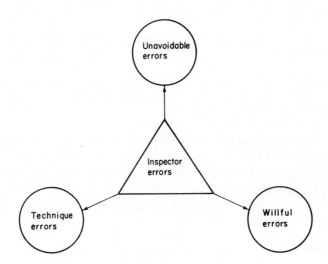

Figure 10.5. Classifications of inspector errors.

Inspector errors classified as technique errors are due to the following causes: (i) lack of "skill" for the job, and (ii) lack of capacity for the job (for example, inadequate training or education, or color blindness).

Another category of inspector errors are known as unavoidable or inadvertent errors. This classification includes those errors that human beings (inspectors) cannot avoid even though they work with the best intentions. Obviously, it is impossible to achieve human perfection.

Finally, the third category of inspector errors are known as willful errors. In this case the inspector has the knowledge that he is making the error and still intends to do it. These can be classified into two separate categories: inspector-initiated and management-initiated willful errors. Some of the forms of inspector-initiated errors are inspector fraud, rounding off and inspector shortcuts. Similarly, the forms of management-initiated errors are management fraud, management apathy, and conflicting management priorities. All of these forms lead to willful inspector errors.

INSPECTION-RELATED MATHEMATICAL MODELS

This section presents four mathematical models concerned with inspection performance. These models follow.

Model I

This model was developed by Juran in 1935 [6] to determine the accuracy of an inspector. This model is based on the reasoning that inspectors can accept bad items and reject good ones. Additionally, the work of regular inspectors is reexamined by check inspectors. Thus the accuracy of an inspector in percentage is given by

$$A = \frac{(D - g)}{D - g + d} \times 100 \ , \tag{10.1}$$

where A is the percentage of defects correctly identified by the inspector, g is the number of good items rejected by the regular inspector, d is the number of defects missed by the regular inspector, and D is the number of defects discovered by the regular inspector.

Dividing the top and bottom parts of Eq. (10.1) by D, we get

$$A = \frac{1 - g/D}{1 - g/D + d/D} \times 100 \ . \tag{10.2}$$

Letting $\theta_1 = g/D$ and $\theta_2 = d/D$ in Eq. (10.2) results in

$$A = \frac{1 - \theta_1}{1 - \theta_1 + \theta_2} \ . \tag{10.3}$$

The plots of Eq. (10.3) are shown in Fig. 10.6. These plots show that as the values of θ_1 and θ_2 increase the inspector accuracy decreases accordingly.

EXAMPLE 10.1

A lot of items were inspected by an inspector, who found 100 defects. A check inspector was assigned to reexamine the entire lot. The check inspector found

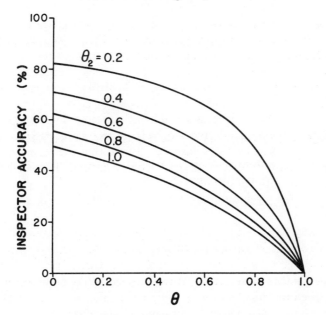

Figure 10.6. Plots of inspector accuracy.

that the regular inspector missed 20 defects and rejected 10 good items. Compute the accuracy of the regular inspector.

In this example the specified values of D, g and d are 100, 10 and 20, respectively. Substituting these values in Eq. (10.1) yields

$$A = \frac{100 - 10}{100 - 10 + 20} \times 100$$

$$= 81.82\% \ .$$

Thus the accuracy of inspector is 81.82%.

Model II

This mathematical model is used to compute waste produced by an inspector (i.e., the percentage of good items rejected by the regular inspector). Thus the percentage of defect-free items rejected by the regular inspector is given by

$$D_f = \frac{g}{T - D - d + g} \times 100 \ , \tag{10.4}$$

where D_f is the percentage of good items rejected by the regular inspector and T is the total number of items inspected. Remaining symbols used in Eq. (10.4) are the same as for model II.

EXAMPLE 10.2

A lot containing 200 items was inspected by an inspector. The inspector found that 80 of the items were defective. A check inspector was assigned to reexamine the entire lot. The check inspector found that the regular inspector missed 10 defects and rejected 5 good items. Calculate the percentage of defect-free items rejected by the regular inspector.

Substituting the given data into Eq. (10.4) results in

$$D_f = \frac{5}{200 - 80 - 10 + 5} \times 100 = 4.35\% \ .$$

Thus the percentage of defect-free items rejected by the regular inspector is 4.35%. If the inspector had rejected 9 good items, the value of D_f would have increased to 7.56%. On the other hand, if he had rejected only one good item, the value of D_f would have been reduced to 0.9%.

Model III

This mathematical model is concerned with obtaining the values of the following four measures of inspection performance [7].

1. The probability of a good item being accepted.
2. The probability of a nonconforming item being rejected.
3. The probability of an inspector rejecting an item.
4. The probability of an inspector accepting an item.

In the first case, the probability of accepting a conforming item is given by

$$P_c = \frac{I_a}{I_a + I_r} \ , \tag{10.5}$$

where I_a is the number of conforming items accepted, I_r is the number of conforming items rejected, and P_c is the probability of accepting a conforming item. Similarly, in the second case the probability of a nonconforming item being rejected is given by

$$P_{nc} = \frac{I_{nr}}{I_{nr} + I_{na}} \ , \tag{10.6}$$

where P_{nc} is the probability of a nonconforming item being rejected, I_{nr} is the number of nonconforming items rejected, and I_{na} is the number of nonconforming items accepted.

In the third case the probability of an inspector rejecting an item [8] is given by

$$P_{ir} = (1 - P_c)(1 - P_r) + P_{nc}P_r \ , \tag{10.7}$$

where P_{ir} is the probability of an inspector rejecting an item, and P_r is the probability of a nonconforming item reaching the inspection stage.

Finally, the probability of an inspector accepting an item is given by

$$P_{ia} = P_c(1 - P_r) + (1 - P_{nc})P_r \ , \tag{10.8}$$

where P_{ia} is the probability of an inspector accepting an item.

Model IV

This model is concerned with obtaining the probability, P_d, of detection of an imperfection in a time t. This probability is given by [8]

$$P_d = 1 - e^{-(1/T)t} \ , \tag{10.9}$$

$$T \equiv \frac{T_m \alpha}{A_v IP} \ . \tag{10.10}$$

where T is the average search time, P is the probability that an imperfection will be detected if it is fixated, T_m is the mean time for a single fixation, I is the total number of imperfections on the object, A_v is the area of the visual lobe, and α is the area of object searched.

SUMMARY

This chapter briefly discusses various aspects of human factors in quality control. Factors influencing the performance of a person are organizational, physical and individual factors. A number of human element considerations in quality assurance are briefly described. Two types of human errors are explained; these are management-controllable and operator-controllable errors.

Five functions of quality supervisors are discussed. These are developing plans, making decisions, communication, personnel development, and monitoring and assessing performance. Various aspects of inspection are covered: inspection objectives, inspection tasks, factors influencing the accuracy of inspection, ways to improve inspection performance, information needed for written inspection instructions, illumination and types of inspection errors. Finally, the chapter presents four inspection-related mathematical models.

EXERCISES

1. A lot was inspected by an inspector, who found 400 defects. The entire lot was reexamined by the check inspector, who found that the regular inspector missed 30 defects and rejected 20 good items. Calculate the accuracy of the regular inspector.
2. What are the factors that influence a person performing a quality function?
3. Explain the following two terms:
   a. Management-controllable errors,
   b. operator-controllable errors.
4. What are the main functions of a quality supervisor?
5. What are the advantages and disadvantages of written and oral communications?
6. List at least eight objectives of inspection.
7. What are the factors that influence the accuracy of inspection?
8. What techniques are useful for improving inspection performance?
9. Describe the following types of inspector errors with examples:
   a. technique errors,
   b. willful errors,
   c. inadvertent errors.

REFERENCES

1. L. W. Rook, *Motivation and Human Error.* Report SC-TM-65-135. Sandia Laboratories, Albuquerque (1965).
2. D. H. Harris and F. B. Chaney, *Human Factors in Quality Assurance.* John Wiley & Sons, New York (1969).
3. C. C. Erhardt, Human factors in quality control, in *Proceedings of the Annual Conference of the American Society for Quality Control,* American Society for Quality Control, Milwaukee, Wisconsin, pp. 446–452 (1981).
4. J. M. Juran, *Quality Control Handbook.* McGraw-Hill, New York (1974).
5. J. M. Juran and F. M. Gryna, *Quality Planning and Analysis.* McGraw-Hill, New York (1980).
6. J. M. Juran, Inspectors' errors in quality control. *Mechanical Engineering* **57**, 643–644 (1935).
7. C. G. Drury, Integrating human factors models into statistical quality control. *Human Factors* **20**, 561–572 (1978).
8. C. G. Drury, Improving inspection performance, in *Handbook of Industrial Engineering* (Edited by G. Salvendy), pp. 8.4.1–8.4.14. John Wiley & Sons, New York (1982).

Chapter 11

Human Factors in Design

INTRODUCTION

Human welfare is affected in many ways by engineering products. People may be users, maintainers, or operators of the designed product and may benefit directly from it. An important role is therefore played by the human factors engineer during system design. The human factors engineer plays dual roles during system development. In one of these roles he or she acts as a potential user, and examines the system from aspects such as human safety, comfort and ease of operating the equipment. In the other role the human factors engineer examines people as elements of the system and determines their overall contribution to the system. Broadly speaking, the human factors engineer plays a vital role in assuring that the design engineers are aware of human needs, desires and limitations.

Some of the objectives of the human factors program in design are [1]

1. to reduce losses from accidents and misuse,
2. to improve user acceptance,
3. to improve human performance and manpower utilization,
4. to reduce the costs of training,
5. to improve the economy of maintenance, and
6. to improve human reliability.

This chapter discusses various aspects of human factors in design.

EFFECTIVENESS OF HUMAN FACTORS CONSIDERATIONS IN ENGINEERING DESIGN

Various studies indicate the effectiveness of human factors considerations at the equipment design stage. The majority of personnel performance errors when using complex equipment are design induced. With proper consideration of human aspects during design, these errors can be significantly reduced. This is clearly demonstrated by Fig. 11.1. This diagram is based on the result of a study conducted by the United States Navy [2]. The study was associated with some complex Navy mine test equipment. Figure 11.1 shows that the total number of errors committed was significantly reduced with human-engineered design as opposed to standard design.

HUMAN FACTORS CONSIDERATIONS IN VARIOUS STAGES OF A SYSTEM

This section presents in the form of questions the human factors considerations in the four phases of a system. These phases are [2]

1. preliminary design phase
2. advanced design phase
3. mock-up to prototype fabrication phase
4. test and evaluation phase.

During the preliminary design phase the human factors engineer seeks answers to questions such as the following.

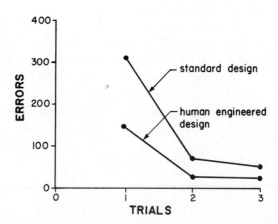

Figure 11.1. Comparison of the total number of errors committed with complex Navy mine test equipment.

1. Who are the users of the system under design?
2. What are the environments in which the system is expected to operate?
3. What is the function that the system under consideration is expected to accomplish?
4. Who will operate and maintain the system?
5. What are the important technological options?
6. What are the mission execution stages?

Similarly, during the advanced design phase the human factors engineer seeks answers to questions of following types.

1. What information is needed by operators and maintenance personnel to accomplish their functions effectively?
2. What kind of indicators and other aids are needed to eliminate operator errors?
3. What type of special skill is expected from operators for effective performance?
4. What is the size of the manpower required during normal operation?
5. What is the size of the manpower required during peak load conditions?
6. What means are available for quick recovery when operator error occurs?
7. What are the conditions that will be detrimental to operator performance?
8. What functions are to be assigned to operating personnel?
9. In what form will the operator find the information most useful to perform his or her task effectively?

In the mock-up to prototype fabrication phase the human factors engineer spends much effort to find answers to questions such as those below.

1. What effect will any proposed changes in configuration have on safety, human performance, reliability, etc.?
2. What quality control methods and procedures will fulfill the task of ensuring the validity of human factors considerations in the resulting system?
3. Are there any options for simplifying any of the instrumentation used in the product?

In the last phase, known as the test and evaluation phase, the human factors engineer is concerned with questions such as how the testing and evaluation can be made as realistic as possible with respect to operating personnel, and what criteria of equipment and operating personnel performance are logical with respect to functions and assigned mission.

USEFUL GUIDELINES FOR DESIGNERS WITH RESPECT TO SELECTED HUMAN FACTORS

This section briefly presents guidelines for designers in the following areas [3]:

1. tackling human factors problems
2. visual and auditory channels
3. body size
4. aging and biorhythms
5. man as a system component.

Guidelines for each of the above areas are discussed below. Some of the guidelines the designer should follow when tackling human factors problems follow.

1. Identify all important product system features correctly.
2. Examine similar systems in operation.
3. Converse with potential users of the system under design.
4. Develop a specification for human-related problems.
5. Ensure that the subjects are truly representative of potential product users.
6. Make sure that the measurements taken of the subjects are going to fulfill the requirements effectively.

The designer will find the following guidelines quite useful when dealing with visual and auditory channels.

Visual Channel

This type of channel for information is most useful in situations such as follows, where

1. environments are noisy,
2. the position of the operator is fixed,
3. messages are long or complex, and
4. the message contains spatial information.

Auditory Channel

This type of channel for information is most suitable in conditions such as follows, where

1. environments are not very good for visual displays,
2. messages are simple or short,
3. the operator is mobile, and
4. a fast response requirement is a must.

Another set of guidelines for designers is concerned with body size. Some of these guidelines follow below.

1. Aim to accommodate 90–95% of the user population.
2. Take into consideration the variation in human dimensions from one country to another when calculating the dimensions of a work space.

3. Design in such a way so that variation of posture is permitted.
4. Consider postural and size problems together.
5. Take into consideration any effects of clothing.
6. Identify displays and controls required by system users.
7. Identify all potential users of the design under consideration.

Some useful guidelines for designers concerning aging and biorhythms follow.

1. Avoid recruiting workers for the night shift who are over 40 years of age.
2. Avoid acquiring habits from older manpower that are contrary to habits already established.
3. Technological changes will adversely affect older workers.
4. Performance will be maintained by older workers because of past experience if they work on a familiar task.
5. It is quite likely that performance will be affected by age when considering speedier and more complex motor tasks.

Finally, with respect to man as a systems component, the following guidelines are useful for designers.

1. A worker's performance may degrade with respect to time on the job.
2. A worker must not be forced to function near his or her maximum load limit for a too long a period.
3. Typically a person needs more than 250 ms to respond.
4. In the sound spectrum a worker's sensitivity range is limited to 20–20,000 Hz.
5. The input capacity of man is easily saturated.

HUMAN SENSORY CAPACITIES AND HUMAN REACTIONS TO EXTREME ENVIRONMENTS

This section briefly describes the above two areas. Four human sensory capacities with respect to the four items shown in Fig. 11.2 are discussed.

In addition, this section discusses human reactions to extreme temperatures such as heat, cold and windchill. Understanding of these areas is important to designers in order to produce effective design [4]. Furthermore, these areas may also affect human reliability. Human sensory capacities and human reactions to extreme environments are briefly described below.

Noise

A human reacts in various ways to noise; fatigue, boredom, and feelings such as well-being, for example. The noise may affect tasks requiring intense concentration or a high degree of muscular management. Furthermore, exces-

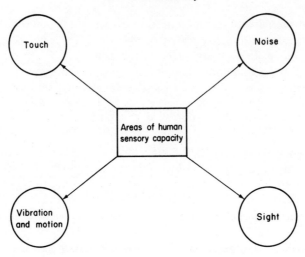

Figure 11.2. Areas of human sensory capacity.

sive noise may damage hearing or make oral communication between persons ineffective.

A noise level below 90 dB is considered to be harmless to human beings. However, with noise levels above 130 dB human beings may experience pain. In any case, noise levels above 100 dB are considered unsafe.

Sight

The human eye sees differently from different angles. A person looking straight ahead can perceive all colors. The designer should pay attention to the following.

1. A person's eyes during the day are most sensitive to greenish-yellow light with a wavelength of about 5500 Å.
2. An increase in viewing angle leads to a decrease in color perception.
3. Color makes little difference in the dark.
4. In the event that critical tasks have to be performed by fatigued persons, it is unwise to place too much reliance on color.
5. Warning lights should be as close to red in color as practicable.
6. It is wise to use red filters, whenever permissible, with wavelengths longer than 6500 Å.
7. Researchers have found that when a person stares at a red or green light and then glances away from it, the color signals may be reversed by the brain. This phenomenon may cause accidents.

Vibration and Motion

The mental and physical task performance of a person may be degraded by vibrations. Vibrations of large amplitude and low frequency contribute to factors such as

1. fatigue and eye strain,
2. interference with interpretation of instruments,
3. motion sickness and headaches, or
4. interference with the ability to read.

Furthermore, low-amplitude and high-frequency vibrations contribute to fatigue.

Touch

This is an important human sense. Touch helps to relieve the load on human eyes and ears by conveying messages to the brain. One example of this important human sense is that a person can recognize different control knob shapes by touching. It is important that designers make use of touch whenever necessary.

ILLUMINATION

The illumination aspect also plays a contributory role in human reliability. Poor workspace illumination may lead to various kinds of human errors. Therefore, this section presents two aspects of workplace illumination [1].

Light Distribution

This section describes the following three methods used for artificial light distribution over the task area: (i) direct light, (ii) indirect light, and (iii) diffused light.

In the case of direct light the rays from the light source fall directly on the work area where the task is being performed. This is achieved with a light bulb with an opaque bowl inverted over it. Between 90 and 100% of the output of the luminaire is directed downward to the working surface. Some of the disadvantages of the direct light method are (i) shadows, (ii) glare and (iii) contrasts.

In indirect lighting most of the rays from the light are reflected from the ceiling and walls before they hit the place of work. An opaque bowl under the light is used to obtain indirect light. With this method close to 90% or more light is directed upward, i.e., toward the ceiling and the upper portion of the surrounding walls. One of the advantages of this method is that it reduces visual fatigue.

Finally, in diffused lighting the light is emitted from a bigger surface area because the light source is enclosed in a translucent bowl. The main advantage of this method is that it consumes less electricity in comparison to indirect lighting. However, it creates some shadows and glare.

Reflection Hazards

Reflection can be very hazardous. For example, a reflection reaching an operator's eyes from a windshield reduces his or her ability to look out [1]. This may lead to catastrophic results. Therefore, the designer must take proper measures to ensure that reflections are not hazardous. Some of these measures are shown in Fig. 11.3.

PROCEDURES AND PROCEDURE EVALUATION

Procedures are used to describe equipment operation, maintenance, etc. For example, individual operator procedures outline how to perform operating and maintenance tasks associated with a system. Proper care is necessary when designing such procedures. Poorly written or designed procedures may result in reduced efficiency of system use and human errors.

According to Ref. [5], information such as that given below should be included in a procedure:

1. equipment descriptions
2. setting up and securing operations

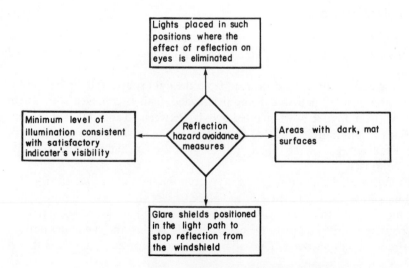

Figure 11.3. Reflection hazard avoidance measures.

3. safety precautions
4. timed instructions sequences
5. persons needed to carry out the task
6. capitalization of critical operations.

Causes of Errors

There are various procedural factors that may lead to error. These factors are associated with

1. discrimination
2. inadequate input characteristics
3. speed and precision
4. coordination.

We first note that procedural factors associated with discrimination requirements that may lead to error are display monitoring over prolonged intervals, making rapid comparison of two or more displays, and basing decisions on multiple source inputs.

Second, procedural factors involved with discrimination requirements that may lead to error are (i) unsatisfactory outlining of visual feedback, (ii) displays requiring discrimination that change quickly, and (iii) having many common characteristics that are associated with displays requiring discrimination.

Third, the procedural factors associated with speed and precision requirements that may lead to error are short decision-making times, and the presence of steps to be performed at high speeds and at very precise times.

Finally, procedural factors involved with coordination requirements that may lead to error are having more than one operator perform between steps, between high-speed control manipulations, or on separate equipment consoles.

PRODUCT SAFETY CONSIDERATIONS IN THE DESIGN PROCESS

Any designed product must be safe to use and operate. During the design process the safety of the product must be examined from several different aspects. In Refs. [6,7] a seven-step procedure that is to be included in the design process is given. This approach will be useful for producing a reasonably safe designed product. Steps of the approach are outlined in Fig. 11.4.

SELECTED USEFUL QUESTIONS FOR DESIGNERS

This section presents a list of selected questions with respect to human factors. These questions should serve as a good start in the design process. The questions are [7]:

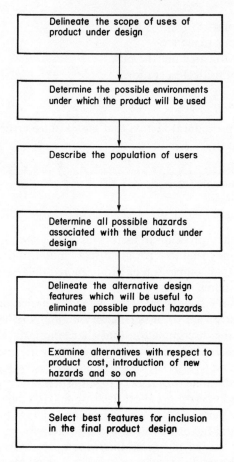

Figure 11.4. Steps for designing a reasonably safe product.

1. What are the functions that need to be performed by humans?
2. What type of displays to transmit information effectively are necessary?
3. Are the displays arranged to obtain optimum results?
4. Are the control devices distinguishable without inconvenience and arranged for optimum results?
5. Are size, shape and other relevant factors being given consideration in the design of controls?
6. Are the proper human factors being given consideration in workspace design?
7. Is proper consideration being given to visibility from the workstation (if applicable) and is it adequate?

8. Are decisions necessarily made by human operators within their capability?
9. Are proper human safety measures considered?
10. Is proper attention given to see that work aids and training complement each other?
11. Are tasks that need to be performed properly grouped into jobs?
12. Are the specified operational requirements of a given control within safe limits?

ADVANTAGES OF HUMAN FACTORS CONSIDERATIONS IN SYSTEM DESIGN

Some of the advantages of human engineering applications in system design are as follows [2].

1. It helps to reduce potential human errors.
2. It can increase system safety.
3. It helps to reduce difficulties in learning equipment operation.
4. It can increase productivity.
5. It helps to reduce difficulties in learning system maintenance.
6. It helps to reduce the cost of personnel training and selection.
7. It helps to reduce equipment downtime.
8. It can reduce operator fatigue.
9. It can provide greater comfort to operators.
10. It is useful for reducing the occurrence of accidents and injuries.
11. It helps to improve user acceptance.

SUMMARY

This chapter briefly discusses the various aspects of human engineering in design. Objectives of human factors engineering are outlined. The effectiveness of human engineering considerations in system design are presented with the aid of a diagram.

Human factor considerations in preliminary design, advanced design, mock-up to prototype fabrication and test and evaluation phases are discussed in detail.

The chapter presents useful guidelines for designers with respect to tackling human factors problems such as visual and auditory channels, body size, aging, biorhythms and man as a system component. Four human sensory capacities with respect to noise, touch, sight, and vibration and motion are explained. Human reactions to extreme environments are briefly discussed.

Another aspect of human factors in design that is covered is concerned with workspace illumination. This aspect covers reflection hazards and light distribution.

Various procedural factors that may lead to errors are presented along with a seven-step product safety procedure for the design process. Finally, the chapter presents selected human-engineering-related questions for system designers and the advantages of human factors considerations in system design.

EXERCISES

1. What are the benefits and drawbacks of human engineering considerations in engineering design?
2. What are the general human-related considerations in the advanced design phase?
3. What are the conditions under which auditory and visual information channels are most suitable?
4. What are the important points associated with human sight?
5. Describe three methods used for artificial light distribution over the task area.
6. Describe the reflection hazard avoidance measures to be considered in system design.
7. What are the procedural factors that may lead to errors?
8. What are the objectives of human engineering in system design?

REFERENCES

1. Joint Army–Navy–Air Force Steering Committee Report, *Human Engineering Guide to Equipment Design*. John Wiley & Sons, New York (1972).
2. H. E. Price, A human factors perspective, in *Proceedings of the Workshop on the Man-Machine Interface and Human Reliability: An Assessment and Projection*, IEEE, New York, pp. 66–67 (1982).
3. J. Wood, Introduction to ergonomics, in *Industrial Design in Engineering: A Marriage of Techniques* (Edited by C. H. Flurscheim), pp. 19–53. Springer-Verlag, New York (1983).
4. *Maintenance Engineering Techniques*. Engineering Design Handbook, pp. 3.68–3.71. AMCP 706-132, produced by Headquarters, U.S. Army Material Command, 5001 Eisenhower Ave., Alexandria, VA 22333.
5. D. Meister and G. F. Rabideau, *Human Factors Evaluation in System Development*, pp. 108–115. John Wiley & Sons, New York (1965).
6. A. Weinstein, A. Twerski, H. Piehler and W. Donaher, *Product Liability and the Reasonably Safe Product*. John Wiley & Sons, New York (1978).
7. E. J. McCormick and M. S. Sanders, *Human Factors in Engineering and Designs*, pp. 504–505. McGraw-Hill, New York (1982).

Chapter 12

Mathematical Models

INTRODUCTION

In engineering disciplines mathematics plays an instrumental role. Various kinds of mathematical concepts are used to solve day-to-day engineering problems. Therefore, it is not wrong to state that mathematical models and formulas are the basic tools of an engineer. A human factors specialist or human reliability specialist also makes use of these tools because mathematical models and formulas are applicable to various areas of human factors engineering. Examples of these areas are

1. human reliability prediction
2. design of instrumental displays
3. human energy expenditure in physical activities
4. illumination
5. air traffic control monitoring
6. modeling human–computer interactions.

The main intent of this chapter is to present selected mathematical models and formulas used in human engineering work. Most of the models and formulas presented below will be directly or indirectly useful in human reliability work.

USEFUL FORMULAS

This section presents selected formulas for making various kinds of human engineering decisions.

Formula I

This formula is concerned with calculating brightness contrast. The brightness contrast, B_c, is defined as follows:

$$B_c = \frac{(L_b - L_d)(100)}{L_b} \, , \tag{12.1}$$

where L_b is the luminance of the brighter of two contrasting areas, and L_d is the luminance of the darker of two contrasting areas.

EXAMPLE 12.1

A specific type of paper has a reflectance of 87%. If the print on the paper has a reflectance of, say, 12%, compute the value of the B_c with the aid of Eq. (12.1).

In this example the specified values of L_b and L_d are 87% and 12%, respectively. Thus from Eq. (12.1) we get

$$B_c = \frac{(87 - 12)(100)}{87} = 86.21\% \, .$$

The value of the brightness contrast is 86.21%.

Formula II

This formula is concerned with evaluating the minimum required display size for identification of a target. The size of a display along an axis with a given range on ground being displayed is given by [1]

$$S_d = (d_v r_g A_{tv})(1.54)/D_t \, , \tag{12.2}$$

where d_v is the viewing distance of the display in inches; r_g is the range, in statute miles, on the ground being displayed; A_{tv} is the minimum target visual angle for detection defined in minutes of arc; D_t is the maximum target dimension given in feet; and S_d is the size of the display, in inches, along the axis on which r_g is being displayed.

In the event that no information regarding the minimum target visual angle value is available, it is advisable to use the value of 12 min. In addition, the minimum value of the display viewing distance, d_v, is usually 16 in.

Formula III

This formula is concerned with computing the recognition distance for steady red and green signal lights [1]. The formula yields an approximation for clear air and daylight situations. The distance, d_f, in feet is given by

$$d_f = c\lambda \, , \qquad (12.3)$$

where $c = 2000$ and λ is the intensity for a similar unit with a clear lens, in candles. For flashing signal lights, the following formula should be used for intensity in Eq. (12.3):

$$\lambda_e = \lambda\tau/(\tau + 0.09) \, , \qquad (12.4)$$

where τ is the duration of flash given in seconds, λ_e is the effective intensity in candles, and λ is the steady light intensity.

The assumptions associated with Eq. (12.4) are outlined in Refs. [1,2].

Formula IV

This formula is used to compute the probability of looking from instrument X to instrument Y and vice versa. This probability is also known as the link value between any two instruments or the transitional probability. According to Refs. [1,3], the transitional probability can be evaluated on the basis of the individual instrument fixation probabilities. The transitional probability between instruments X and Y is given by

$$P_{XY} = (2P_X P_Y) \bigg/ \left(1 - \sum_{j=1} P_j^2\right) \, , \qquad (12.5)$$

where P_j is the jth instrument fixation probability, P_X is the probability of being on instrument X, and P_Y is the probability of being on instrument Y. With this model attention should be paid to the following points.

1. The most frequently fixated-upon instrument should be placed in the center of panel under study.
2. The instruments with the lowest link values should be placed farthest from the center.
3. The instruments with high link values should be placed peripherally adjacent to the central instruments.

Formula V

This formula is used to find the value of the visual angle. The importance of the visual angle stems from the fact that the message legibility is subject to the visual angle subtended at the person's eye by the viewed object [4]. The constant values given in this formula can only be used for angles less than $10°$. The visual angle, θ, in minutes is given by

$$\theta = (60(57.3)h_d)/D \, , \qquad (12.6)$$

where D is the distance from the target and h_d is target diameter or the height of the character.

Formula VI

This formula is concerned with calculating the value of the control–display (c/d) ratio. This is defined as the ratio of the control movement distance to that of the display moving element. It is applicable only to continuous controls. This ratio is considered to be an important design factor affecting the performance of operators. Various studies have indicated that in comparison to a poor control–display ratio, a good control–display ratio can save from 0.5 to 5 s in positioning time. The following control–display ratio formula is presented in Ref. [1] for controls that involve considerable rotational movement and that affect linear displays:

$$(c/d) = [2\pi\beta\alpha(360)^{-1}]/d_m \ , \tag{12.7}$$

where β is the lever arm length, d_m is the display movement and α is the control angular movement given in degrees.

The range of optimum values for the control–display ratios are as follows: (i) 0.2–0.8 for knobs, (ii) 2.5:1 to 4:1 for ball levers.

Formula VII

This formula is used to calculate the optimum character height. The formula becomes useful when labeling panel fronts. The formula is defined in terms of factors such as (i) viewing conditions, (ii) the importance of numbers, and (iii) the viewing distance. According to Refs. [5,6], the height, h_ℓ, of a letter in inches is given by

$$h_\ell = (0.0022)d + \sum_{i=1}^{2} c_i \ , \tag{12.8}$$

where c_i is the ith correction factor for $i = 1$ (correction factor for viewing condition), $i = 2$ (correction factor for the criticality of the number), and d is the viewing distance defined in inches. Various values for c_1 and c_2 are given in Table 12.1.

Formula VIII

This is concerned with computing the value of illuminance or illumination. The illuminance is measured in terms of luminous flux per unit area. The lumininous flux is the rate at which light is emitted from a source. Thus the illuminance is defined [7] by

Table 12.1. Values for correction factors c_1 and c_2

No.	Condition	Value of c_1	Value of c_2
1	Unfavourable reading conditions; high ambient illumination	0.16	—
2	Unfavourable reading conditions; low ambient illumination	0.26	—
3	Favourable reading conditions; low ambient illumination	0.16	—
4	Favourable reading conditions; high ambient illumination	0.06	—
5	The number is very critical	—	0.075
6	The number is other than very critical	—	0

$$\mathrm{IL} = \frac{L_i}{d_s} , \qquad (12.9)$$

where IL is the illumination in lux (a lux is one lumen per square meter), L_i is the luminous intensity in candelas, and d_s is the distance, in meters, from the light source.

Formula IX

This formula is used to calculate the value of reflectance. The reflectance, R, is given by

$$R = \frac{L_m}{\mathrm{IL}} , \qquad (12.10)$$

where L_m is the luminance in candelas per square meter (the luminance is the amount of light per unit area departing from a surface). It is to be noted that for a perfect reflecting surface the value of the reflectance is equal to 1.

Formula X

This formula is concerned with computing the value of the glare constant. According to Ref. [6], the most modern formula to measure the value of the glare constant, g_c, is defined as follows:

$$g_c = A_s^{\alpha_1} S_\ell^{\alpha_2} / L_{gb} A_{vg}^2 , \qquad (12.11)$$

where $\alpha_1 \equiv 0.8$, $\alpha_2 \equiv 1.6$, A_s is the solid angle subtended by the source at the eye, S_ℓ is the source luminance, A_{vg} is the angle between the glare source direction and the viewing direction, and L_{gb} is the luminance of the general background.

A value of the glare constant equal to 35 indicates the boundary of "just acceptable" glare, and 150 the boundary of "just uncomfortable" glare.

Formula XI

This formula is used to determine the human energy costs associated with lifting weights [8]. The energy cost, C_E, in kilocalories per hour is defined as

$$C_E = \frac{WNCH}{1000} \, , \tag{12.12}$$

where W is the weight in pounds, N is the number of lifts in a 1 hr period, C is the cost of energy per lift (gcal/ft·lb), and H is the lifting height in feet.

EXAMPLE 12.2

Calculate the value of the energy cost, C_E, if the values of W, N, C and H are 12 lb, 150 lifts/hr, 4 gcal/ft·lb and 2 ft, respectively.

With the aid of given data and Eq. (12.12), we get

$$C_E = \frac{WNCH}{1000} = \frac{(12)(150)(4)(2)}{1000}$$

$$= 14.4 \text{ kcal/hr} \, .$$

Formula XII

This formula is used to compute the total amount of rest (scheduled or unscheduled) required for any specified work activity [8]. The total rest, R_t, required in minutes is given by

$$R_t = t_w(C_a - C_s)/(C_a - 1.5) \, , \tag{12.13}$$

where C_s is the level of energy expenditure, in kcal/min, adopted as standard; C_a is the average kilocalories expenditure per minute of work; and t_w is the total amount of working time given in minutes.

For an "average" person the value of C_s may be taken as 5 kcal/min. The constant value of 1.5 in Eq. (12.13) represents the approximate resting level in kcal/min.

EXAMPLE 12.3

In Eq. (12.13), the specified values of t_w, C_a and C_s are 30 min, 10 kcal/min and 5 kcal/min, respectively. Compute the value of the total rest period required in minutes.

Substituting the specified data in Eq. (12.13) results in

$$R_t = \frac{30(10 - 5)}{10 - 1.5} = 17.65 \text{ min} .$$

The value of the total rest period is 17.65 min.

Formula XIII

This formula is concerned with calculating the maximum lifting load for males and females [9]. In the case of males the maximum lifting load, L_m, is given by

$$L_m = C_m S_m , \tag{12.14}$$

where $C_m = 1.1$ and S_m is the isometric back muscle strength.

Similarly, for females the maximum lifting load, L_{mf}, is given by

$$L_{mf} = C_f S_f , \tag{12.15}$$

where $C_f = 0.95$ and $S_f =$ the isometric back muscle strength.

Formula XIV

This formula is used to calculate the "reading ease score for a 100-word sample passage of the prose." The formula is according to Flesch [10]. Thus, the reading ease score, S_{re}, is given by

$$S_{re} = i - jS - kn_w , \tag{12.16}$$

where $i = 206.835$, $j = 0.84$, $k = 1.015$, S is the number of syllables in given 100 words, and n_w is the mean number of words per sentence. A value of S_{re} equal to 100 indicates easy to ready for any literate person. A zero score indicates practically unreadable material.

Formula XV

This formula is used to determine sound pressure levels. The sound power is directly proportional to the square of the sound pressure. The sound pressure level, $S_{p\ell}$, in decibels (dB) is given by

$$S_{p\ell} = 10 \log(P_1/P_0)^2 \ , \tag{12.17}$$

where P_1 is the sound pressure to be measured, and P_0 is the standard reference sound pressure (the value of P_0 cannot be equal to zero). The human ear can hear sounds from about 0.0001 to 1000 μbars of sound pressure [1]. The value of 0.0002 μbar is used as the standard reference sound pressure.

Formula XVI

This formula is known as the Shannon formula and is used to compute the channel capacity needed to transmit perfectly intelligible speech [1]. The equation for the formula is defined as follows:

$$C_c = b_c \log_2 \left(1 = \frac{s_p}{n_p} \right) \ , \tag{12.18}$$

where C_c is the channel capacity in bits per second, s_p is the signal power, n_p is the noise power, and b_c is the bandwidth of the channel.

Formula XVII

This formula is concerned with measuring inspector performance associated with inspection tasks. Thus the inspector performance, I_p, is given [11] by

$$I_p = (t_{it})/(m - n_e) \ , \tag{12.19}$$

where n_e is the number of inspector errors, m is the number of patterns inspected, t_{it} is the total inspection time, and I_p is the inspector performance in minutes per correct inspection. It is to be noted that the above index is an average over a specified trial.

Formula XVIII

This formula is known as the maintainability index and was developed by the United States Navy Electronics Laboratory. This index is used to describe the effect of total accumulated maintenance time on the availability of a system. The index is defined below:

$$I_m = \left[1 - \frac{T_{cm}}{\tau + T_m} \right] \times 100 \ , \tag{12.20}$$

where τ is the total system operation time in hours, and T_{cm} is the total amount of corrective maintenance downtime per τ hours of system operation (this time is given in units of hours). The total system operation time in hours is expressed as follows:

$$\tau = T_c - T_{ei} \ , \tag{12.21}$$

where T_c is the calendar time over which the value of T_{cm} is obtained, and T_{ei} is the total time the system is in inoperative condition.

MATHEMATICAL MODELS

This section presents three selected mathematical models developed relatively recently in comparison to the many formulas of the previous section (Useful Formulas).

Model I

This model was developed by Rashevsky [12] and is concerned with determining the maximum speed of a car on a traffic-free straight highway. The maximum safe speed, S_m, to drive is expressed as follows:

$$S_m = (W_h - W_c - 2D_{ms} - L_c A_v)/A_v t_r \ , \tag{12.22}$$

where t_r is the driver reaction time, A_v is the mean angle by which the direction of the vehicle under study sometimes deviates from the actual course, W_h is the highway width, W_c is the car width, D_{ms} is the car's minimum safe distance from the pavement edge, and L_c is the car length.

A formula to compute driver's reaction time is also given in Ref. [12].

Model II

This model is used to estimate the mean risk of an accident given an encounter of a car with child pedestrians. From Ref. [13], the mean risk, R_m, of an accident given an encounter with a car is expressed as follows:

$$R_m = 1 - \frac{1}{\bar{m}_c} \ln (1 - k) \ , \tag{12.23}$$

$$\bar{m}_c \equiv \frac{\sum m_c}{N} \ , \tag{12.24}$$

$$m_c \equiv \sum L_p \ , \tag{12.25}$$

$$L_p \equiv \frac{L_c + S_c t_a}{S_a} \ , \tag{12.26}$$

$$k \equiv \frac{m_a}{N_{cp} d} \ , \tag{12.27}$$

HR-I*

where $\bar{m}_c$ is the average number of cars encountered in a day, m_c is the number of cars a given pedestrian will encounter for all of the roads crossed in a day, N is the number of child pedestrians observed, L_p is the proportion of the road length occupied by moving traffic, L_c is the mean length of cars, S_c is the mean speed of cars, t_a is the mean time taken by a child pedestrian to cross the path of a car, S_a is the mean spacing of cars, m_a is the number of child pedestrian accidents during a specified period for a specific road, N_{cp} is the number of child pedestrians in the population, and d is the number of days over which data for m_a have been collected.

Model III

This model is useful for describing aircraft pilots' behavior under real flight conditions. The model makes use of queueing theory.

Most of the basic assumptions associated with this model [11] are as follows.

1. Instruments compete for the pilot's attention.
2. At each instant the aircraft pilot looks at any one of the instruments, the pilot is postponing the observation of other instruments.
3. The aircraft pilot does his or her best by selecting an instrument for observation so that the risk associated with not observing the other instruments is minimal.
4. Each instrument reading time is constant.
5. Costs are assigned to each instrument.

The total cost of not looking at any instrument at time t is defined as

$$K(t) = \sum_{n=1}^{y} k_n P_n(t)[1 - P_n(t)]^{-1} , \qquad (12.28)$$

where k_n is the cost for exceeding the established threshold value for the nth instrument; P_n is the probability that the nth instrument will bypass its threshold value (i.e., danger mark), g_n, at time t; y is the quantity of instruments; and t is the observation time.

The total cost of looking at the mth instrument at time t is given by

$$K_m(t) = K(t) - K_m P_m(t) . \qquad (12.29)$$

For the maximum value of $K_m P_m(t)$, the value of $K_m(t)$ will be at its minimum. Thus the strategy should be to choose the mth instrument such that it makes the value of $K_m(t)$ minimum. The detailed description of this model may be found in Ref. [14].

SUMMARY

This chapter presents 18 selected formulas and three mathematical models. Formulas are concerned with computing the value of

1. brightness contrast
2. display size
3. recognition distance for steady red and green signal lights
4. probability of looking from instrument X to instrument Y
5. visual angle
6. control–display ratio
7. optimum character height
8. illumination
9. reflectance
10. glare constant
11. human energy cost associated with lifting weights
12. rest period
13. maximum lifting load for males and females
14. reading ease score
15. sound pressure level
16. channel capacity
17. inspector performance
18. maintainability index.

The three mathematical models are concerned with determining the maximum safe speed to drive a car, the mean risk of an accident given an encounter with a car by child pedestrians, and the behavior of pilots under real flight conditions. Source references for the material presented in this chapter have been noted.

EXERCISES

1. A certain kind of paper has a reflectance of 83%. Assume that the print on the paper has a reflectance of 9%. Determine the value of the brightness contrast.
2. The following values are defined for the symbols used in Eq. (12.12): $W = 10$ lb, $N = 120$ lifts/hr, $C = 5$ gcal/ft·lb, and $H = 18$ in.

   Calculate the value of human energy costs associated with lifting weights. Comment on the calculated value.
3. With the aid of Eq. (12.13) determine the value of the total rest required if $C_s = 4$ kcal/min, $t_w = 0.5$ hr and $C_a = 11$ kcal/min. Compare the calculated value with the total working time.
4. Discuss any three of the following: (i) visual angle, (ii) brightness contrast, (iii) sound pressure level, or (iv) decibels.

5. Explain the meaning of the following two terms: (i) glare, and (ii) discomfort glare.

REFERENCES

1. *Human Engineering Guide to Equipment Design*, sponsored by Joint Army-Navy-Air Force Steering Committee. John Wiley & Sons, New York (1972).
2. J. E. Wesler, The effective intensity of flashing lighted aids to navigation, presented at the *Sixth International Technical Conference on Lighthouses and Other Aides to Navigation*. Report No. AD 242046, U.S. Coast Guard, Washington, D.C. (1960).
3. J. W. Senders, A. re-analysis of the pilot eye-movement data. *IEEE Transactions on Human Factors in Electronics* **7**, 103-106 (1966).
4. W. F. Grethers and C. A. Baker, *Visual Presentation of Information in Human Engineering – Guide to Equipment Design* (Edited by H. P. Van Cott and R. G. Kinkade), Chapter 3. U.S. Government Printing Office, Washington, D.C. (1972).
5. G. A. Peters and B. B. Adams, Three criteria for readable panel markings. *Product Engineering* **30**, 55-57 (1959).
6. D. J. Oborne, *Ergonomics at Work*. John Wiley & Sons, Chichester (1982).
7. E. J. McCormick and M. S. Sanders, *Human Factors in Engineering and Design*. McGraw-Hill, New York (1982).
8. K. F. H. Murrell, *Human Performance in Industry*. Reinhold, New York (1965).
9. E. Poulsen and K. Jorgensen, Back muscle strength, lifting and stooped working postures. *Applied Ergonomics,* **2**, 133-137 (1971).
10. R. Flesch, A new readability yardstick. *Journal of Applied Psychology* **32**, 221-233 (1948).
11. C. G. Drury and J. G. Fox (Eds.), *Human Reliability in Quality Control*. John Wiley & Sons, New York (1975).
12. N. Rashevsky, Mathematical biophysics of automobile driving. *Bulletin of Mathematical Biophysics*, **21**, 375-385 (1959).
13. C. I. Howarth, D. A. Routledge and R. Repetto-Wright, An analysis of road accidents involving child pedestrians. *Ergonomics* **17**, 319-330 (1974).
14. J. R. Carbonnell, A queueing model of many-instrument visual sampling. *IEEE Transactions on Human Factors in Electronics* **7**, 157-164 (1966).

Chapter 13

Applications of Human Factors Engineering

INTRODUCTION

Today human factors engineering is a well-recognized discipline. It is being taught at undergraduate and graduate levels in various universities and similar institutions. Various kinds of research are being conducted in many parts of the world, and results are being transmitted through a number of well-established journals on the field.

Human factors engineering has come a long way since its inception. In those days the applications of human engineering were narrowly confined to specific but important areas such as military systems. Nowadays, the benefits of this discipline are being felt across many areas. Its applications are following a growing trend. Some examples of application areas of human engineering are:

1. nuclear and conventional energy production systems
2. transportation systems
3. engineering systems for national security
4. process control
5. computer-based business information systems.

All of these areas are described in subsequent sections of this chapter.

HUMAN FACTORS IN TRANSPORTATION SYSTEMS

Many new challenges in human factors engineering have been created by the need for improved urban transportation. Here the basic problem for transportation systems designers is the availability of human factors engineering information [1].

The sizing of the vehicle to the user population is an important consideration in passenger vehicle design. The term "sizing" is concerned with (i) storage areas and aisles, (ii) entrances and exits, and (iii) seats. The basic problems of concern with sizing are the existence of necessary strength and anthropometric data, and the correct selection of design criteria.

Human Considerations in Urban Transportation

There are various areas requiring human considerations in urban public transportation. The major areas of consideration [2] are shown in Fig. 13.1. Illustrative problems within each area shown in this figure are discussed briefly below.

Area I. This comprises personal spacing, independence in choosing travel companions, privacy, etc.

Area II. This includes provisions for emergency, risk of accident, crime, maintenance, etc.

Area III. This comprises pollution, compatibility with surroundings and congestion.

Area IV. This includes service frequency, schedule reliability, trip time, messages related to routing schedule, etc.

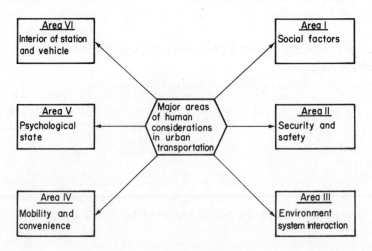

Figure 13.1. Major areas of consideration in urban transportation.

Area V. This comprises the quality of ride, anxiety, boredom, comfort, etc.

Area VI. This includes the quality of air, vibration, noise, seating, illumination, etc.

HUMAN FACTORS IN MILITARY SYSTEMS

Human factors receive significant attention in the development and acquisition of military systems, equipment and facilities in the United States. The military specification MIL-H-46855B entitled *Human Engineering Requirements for Military Systems, Equipment and Facilities* [3] is one example of such attention. This kind of specification serves various purposes [4].

1. It defines the human factors needs applicable at the time when contractors respond to requests for proposals by the military.
2. It helps the military procurement agency to have effective control of the human factors engineering effort.
3. It defines the human factors needs during the research, design and development, test and evaluation phases.
4. It helps in the assessment (with the aid of information provided by the contractor) of the contractor's capability with respect to human factors engineering.
5. It defines the nature and scope of work to be performed by the contractor with respect to human factors.

The United States Army calls for a human factors effort related to the major areas shown in Fig. 13.2. Areas shown in Fig. 13.2 are described in detail in Ref. [4].

Areas of Human Factors Requirement Description in the Proposal

This section briefly outlines the areas of human factors requirement description in the proposal submitted by the contractor in response to a request by the military. With respect to human factors, the proposal should outline a description of the human factors engineering relationship to, and participation in, areas such as follows [4].

1. Structure of the human factors engineering organization. This includes a list of people involved in human factors engineering and their credentials, functions and authority of components of human factors engineering, location of human factors engineering within the overall organizational structure, etc.
2. Environmental considerations analysis with respect to human factors.

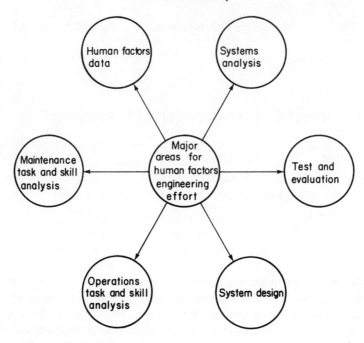

Figure 13.2. Major areas for human factors engineering effort.

3. Anticipation of training and manpower needs. This includes job aids, skill requirements, manning levels, devices for training, etc.
4. Human factors analysis with respect to tasks.
5. Human factors considerations in design review.
6. Integration of human factors into system analysis, design, and trade-off studies.
7. Integration of human factors into testing and evaluation.
8. Anticipation of special studies and mockups required.
9. Human factors integration methods in overall design, human factors data maintenance and human factors integration with quality control, reliability and safety, and so on.

People in the Department of Defense frequently noted major shortcomings with respect to human factors in proposals submitted by contractors. These shortcomings are shown in Fig. 13.3.

HUMAN FACTORS IN NUCLEAR POWER GENERATION

Nuclear power reactors are one of the most complex systems ever built by man. The presence of various kinds of safety systems to prevent and mitigate the occurrence of accidents is an important factor in reactor complex-

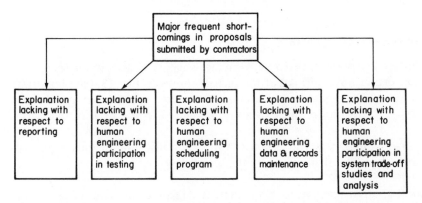

Figure 13.3. Major frequent shortcomings in proposals submitted by contractors.

ity. In the routine operation of a nuclear power station, human beings interact in almost all areas. Therefore human operators are an important factor in the reliability analysis of a nuclear power plant. The subject of human factors in nuclear power generation is wide in nature and can be examined from various different aspects. However, this is not the intent of this section; it concentrates on two areas of human factors in nuclear plants. These are concerned with protection of important controls and performance-shaping factors in nuclear power plant operations. These areas are discussed separately below.

Guidelines for the Protection of Important Controls

This section presents a list of four guidelines [5] for the protection of vital controls. A nuclear power station has a large number of controls. If any of these controls are inadvertently actuated due to a human error, it may lead to very undesirable results (safety injection and reactor trip, for example). Therefore, great care must be taken to protect such controls. The following guidelines will be found useful.

1. To prevent inadvertent actuation, place caution and danger tags on controls in question.
2. To prevent accidental actuation of critical buttons place tall hollow cylinders on them.
3. To prevent operation of equipment that are not to be operated cover their controls with masking tape. Reasons for not using such controls may be written on the back of the tape.
4. To protect vital controls place plastic covers over them. These covers may be held in place magnetically. The presence of such covers will constantly remind the operators of the criticality of such controls.

Performance-Shaping Factors in Nuclear Power Plant Operations

This section presents seven performance-shaping factors in nuclear power plant operations [6,7]. All of the factors shown in Fig. 13.4 are important and must be taken into consideration when deriving estimated human error rates. Each of the factors shown in this figure are briefly discussed below.

Human redundancy. This is concerned with verification of the correct performance of the first person by the second person.

Human actions independence. This term is sometimes also referred to as "coupling of human actions." If there is no coupling between tasks, the probability of error occurrence in one task is independent of the probability of error occurrence in another task. Dissimilar tasks tend to be independent.

Training quality. This is concerned with the quality of training of the nuclear power plant personnel. Examples of such personnel are the operators and maintainers.

Quality of method of use and written material. This is concerned with the quality of the method being used to perform a task as well as the quality of writ-

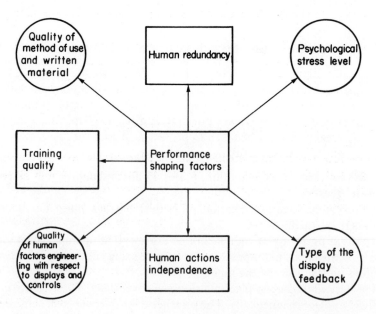

Figure 13.4. Seven performance-shaping factors in nuclear power plant operations.

ten instructions. Factors such as printing quality, procedure format, and ease of understanding determines the quality of the written material.

Quality of human factors engineering with respect to displays and controls. This is concerned with the attention given to human engineering in design with regard to displays and controls; for example, higher error rates are to be assigned where the arrangement and labeling of controls is confusing.

Type of the display feedback. This is concerned with a situation such as too many displays competing for the operator attention at any one time.

Psychological stress level. This aspect is concerned with the performance of nuclear power station personnel under stressful conditions.

HUMAN FACTORS IN PROCESS CONTROL

This section briefly discusses human factors in process control. Since 1960, several studies of operators have been conducted by human factors specialists. One of the significant studies of man and computers in process control was conducted in 1969 [8]. According to Ref. [9], efforts have been made in process control to make use of existing work in areas such as simple tasks, emergency behavior and vigilance tasks. The meanings of each of these three terms are explained below.

Simple task. This is a well-defined sequential operation involving a minimum of decision making.

Vigilance task. This is associated with the detection of a signal. An example of a vigilance task is an operator monitoring the operation and the integrity of the process.

Emergency behavior. This is concerned with human behavior during emergency situations.

Various researchers have already contributed to simple task-related studies in general [9]. However, with respect to process control, a specific approach has been used in Ref. [8]. This study involves the evaluation of the execution reliability of a schedule of trip and warning tests.

Some methods are available for calculating simple and vigilance task reliability in process control. A study related to vigilance in process control was reported in Ref. [10]. This study is concerned with the operators of a nuclear reactor.

Various studies concerning human behavior in emergencies have been conducted. One such study related to process control is reported in Ref. [11]. This study investigates process operator behavior under breakdown conditions.

HUMAN FACTORS IN BUSINESS INFORMATION SYSTEMS

This section discusses human factors in relation to human reliability in computer-based business information systems. Increasing reliance on computer-based business information systems has focused attention on human factors. Various studies associated with keypunching, copying, handprinting, coding and general data transcription surfaced in the 1960s [12]. It may be due to the findings of these studies that designers of business information systems have indicated their preference for applying methods for detecting and correcting human errors rather than for preventing such errors.

Causal Factors Concept

This approach can be used [12] to improve human reliability. It offers new flexibility to system designers as compared to alternatives of the past. These alternatives were either to eliminate or ignore the human component, or to develop a system that tolerated a level of human reliability. Some of the assumptions associated with this approach follow.

1. Human errors can be eliminated by acting on factors that cause them to occur.
2. It is possible to identify and control the human-error-causing factors.
3. A human task will be performed with improved accuracy under optimum conditions.
4. In a situation when one or more causal factors act to degrade performance a human error is generated.

The major causal factors that act or interact to degrade the performance of humans in computer-based business information systems are outlined in Fig. 13.5.

Each of these major causal factors is comprised of several subfactors.

Personal causal factors are perhaps the most complex causes of errors committed by humans. These factors are concerned with psychological and physiological needs (age, etc.). The effect of these factors can only be eliminated if the human functions are performed by machine. In computer-based business information systems personal factors probably account for about 50% of human errors.

Training causal factors are concerned with the training of the workers to perform assigned tasks. If the workers are not fully trained to perform their work, the probability of human error occurrence will be high.

Source data, environmental and man–machine causal factors are probably least important in computer-based business information systems. Source data factors are concerned with those data item characteristics that affect the human ability to process information correctly. Environmental factors are associated with those environments that may affect workers' performance. Examples of such environments are noise, temperature and distraction.

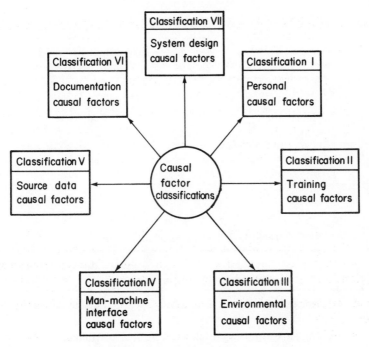

Figure 13.5. Major categories of causal factors.

Man–machine interface factors are associated with man–machine interactions. CRT displays and keyboards are two of the important areas in computer-based business information systems.

The two remaining major categories of causal factors are documentation and system design factors. Documentation factors are associated with documentation used by operators to perform their work. Finally, system design factors are associated with design aspects of the system that may reduce human reliability. An example of a system design subcausal factor is "insufficient time allowed to complete a task."

In computer-based business information systems, the possible breakdown percentage of human errors due to major causal factors is given in Table 13.1. In Table 13.2, examples of important subfactors associated with each of the seven major causal factors are given.

SUMMARY

This chapter briefly discusses the five important areas of human factors engineering applications. These are transportation systems, military systems, nuclear power generation, process control and computer-based business information systems.

Table 13.1. Probable percentage of human errors for major causal factors

Major causal factors	Approximate percentage of human errors
1. Personal	50
2. Training ⎫	
3. Documentation ⎬	40
4. System Design ⎭	
5. Man–machine interface ⎫	
6. Source data ⎬	10
7. Environment ⎭	

Table 13.2. Examples of subfactors

No.	Causal subfactor	Corresponding major causal factor
1	Not enough time to accomplish a task	system design
2	No worker training at all	training
3	Workers cannot understand the documentation	documentation
4	Work area has lighting intensity below 10 foot candles	environmental
5	Characters' brightness contrast to background is 10%	man–machine interface
6	Handprinted characters are not legible	source data
7	Operator has a shortcoming of an important required critical skill	personal

Human considerations in urban public transportation are discussed by outlining the major areas of consideration.

Human factors engineering in military systems plays an important role. The purposes of human engineering specification are listed and major areas for human factors engineering efforts are shown. Areas of human factors requirement description in the proposal are discussed. Major frequent shortcomings in proposals submitted by contractors are outlined.

The next application area is concerned with nuclear power generation.

Guidelines for the protection of important controls are listed, and the seven performance-shaping factors in nuclear power plant operations are described.

Human factors in process control are discussed. Discussion on this topic is centered on simple tasks, emergency behavior and vigilance tasks.

The last topic discussed in the chapter is concerned with computer-based business information systems. Seven major causal factor categories are described. The approximate percentages of human errors associated with these major causal factors are presented.

Examples of causal subfactors associated with each of the major causal factors are tabulated.

EXERCISES

1. Describe the following three terms:
   a. simple tasks
   b. emergency behavior
   c. vigilance tasks.
2. Discuss the causal factor approach used in business information systems.
3. Discuss human factors engineering applications in hospitals.
4. Describe an important human engineering application area that has played an instrumental role in human engineering development.
5. Discuss the present-day trends in applications of human engineering.
6. What are the major areas of human engineering efforts in military systems?
7. What is the meaning of the term "performance-shaping factors"? Discuss two important shaping factors.
8. What are the major factors that will affect human performance in computer-based business information systems?

REFERENCES

1. L. L. Hoag, Human factors in urban transportation systems. *Human Factors* **17**, 119–131 (1975).
2. A. E. Millar (Ed), *The Motion Commotion, Human Factors in Transportation.* NASA-ASEE Report, Contract NGT 47-003-028, Langley Research Center and Old Dominion University Research Foundation (1972).
3. *Human Engineering Requirements for Military Systems, Equipment and Facilities.* MIL-H-46855B, prepared by the Department of Defense, Washington, D.C. (1984). Available from the Naval Publications and Forms Center, 5801 Tabor Ave., Philadelphia, PA 19120.
4. R. F. Chaillet, Human factors requirements for the development of U.S. Army material, in *Proceedings of the Human Operator in Complex Systems* (Edited by W. T. Singleton, R. S. Easterby and D. C. Whitfield). Taylor and Francis, London (1967).
5. G. W. Manz, Human engineering: Aids to smooth operation. *Nuclear Safety* **18**, 223–228 (1977).

6. A. D. Swain and H. E. Guttman, Human reliability analysis applied to nuclear power, in *Proceedings of the Annual Reliability and Maintainability Symposium*, IEEE, New York, pp. 116–119 (1975).

7. *Reactor Safety Study*, Wash-1400 (NUREG-75/014) (October 1975). Available from the National Technical Information Service, Springfield, Virginia 22161.

8. J. F. Ablitt, *A Quantitative Approach to the Evaluation of the Safety Function of Operators on Nuclear Reactors*. Report No. AHSBSR160 (1969). Available from the U.K. Atomic Energy Authority, Risley, Lancashire, England.

9. F. P. Lees, Quantification of man–machine system reliability in process control. *IEEE Transactions on Reliability* **R22**, 124–131 (1973).

10. A. E. Green, *Safety Assessment of Automatic and Manual Protective Systems for Reactors*. Report No. AHSBSR172 (1969). Available from the U.K. Atomic Energy Authority, Risley, Lancashire, England.

11. J. A. Clark, Display for the Chemical Plant Operator. M.Sc. thesis, 1972. Available from the University of Manchester Institute of Science and Technology, Manchester, England.

12. R. W. Bailey, S. T. Demers and A. I. Lebowitz, Human reliability in computer-based business information systems. *IEEE Transactions on Reliability* **22**, 140–147 (1973).

Index

About the Author

Dr. Balbir S. Dhillon is Professor of Mechanical Engineering at the University of Ottawa, Ontario, Canada. He has published over 150 articles on reliability engineering and related areas. Dr. Dhillon is a member of the Editorial Advisory Board of *Microelectronics and Reliability, An International Journal* and is Editor-in-Chief of *Human Reliability: The International Journal*, both published by Pergamon. He served as an Associate Editor of the 10th–13th Annual Modelling and Simulation Proceedings, published by the Instrument Society of America. He also serves as an Editor-at-Large in Engineering for Marcel Dekker, Inc., and as a referee to many national and international journals, including *IEEE Transactions on Reliability*. Recently, Dr. Dhillon has been appointed to the post of Associate Editor for the *International Journal of Energy Systems*. He has written several books on various aspects of system reliability, maintainability, and quality control. His first book on reliability, co-authored by Dr. Singh, was translated and published into Russian by Mir Publishers of Moscow in 1983.

Dr. Dhillon is a recipient of the Society of Reliability Engineer's Merit Award and the American Society for Quality Control's Austin J. Bonis reliability award. He is a registered professional engineer in Ontario with several years of industrial experience. He is listed in the *American Men and Women of Science*, *Dictionary of International Biography*, *Men of Achievement*, and *Who's Who in International Intellectuals*.

Dr. Dhillon attended the University of Wales, where he received his B.Sc. in Electrical and Electronic Engineering, and his M.Sc. in Industrial and Systems Engineering. He received his Ph.D. in Industrial Engineering from the University of Windsor.